Short Takes
Dr. Robert M. Jones

Table of Contents

General Maintenance 5
Bearings 5
- *Analysis of Very Slow Bearings* 6
- *Axial Loads On Rolling Element Bearings* 7
- *Bearing Cage Failures* 8
- *Bearing Clearances* 10
- *Cage Materials* 11
- *Cage Problems* 12
- *"Checking" Bearings* 14
- *How Bearings Are Analyzed Without Removal* 15
- *Determining Damaged Bearings Without ID* 17
- *It Bears Repeating* 19
- *Monitoring Ultra Slow Rolling Element Bearings* 20
- *Overheated Bearings* 21
- *Analysis of Rolling Element Bearings* 23
- *Rolling Element Bearings II* 24
- *Sealed or Open Bearing?* 26
- *Shiny New Bearings* 27
- *Shop Practices: Handling Bearings* 28
- *Unusual Characteristics of Very Slow Bearings* 29
- *Very Slow Bearing Analysis* 30
- *What's the Difference* 33
- *Power* 35

Electrical 37
- *Predictive Maintenance* 38
- *Variable Frequency Drives (VFD) Part I* 39

Gears 40
- *Gear Analysis* 41
- *Gearbox Maintenance* 42
- *Gearbox Trouble Shooting* 44
- *Synchronous Gearbox* 46

Motors 48
- *Adventures in Misalignment* 49
- *Electric Motors and Bearings* 50
- *Electric Motors Part 1* 51
- *Electric Motors Part 2* 53
- *Electric Motors Part 3* 55

Pumps 56
- *Detection of a Rubbing Impeller* 57
- *Misalignment* 59
- *Pipe Bound* 60
- *Pumps and Piping* 61
- *Vertical* 62

Miscellaneous 63
- *Classified Maintenance Service* 64
- *Consequences of Buying Cheap Products* 65
- *Duct Tape and Locktite* 66
- *Equipment Storage* 67

Free Information *68*
If You Don't Believe Us, We're Outta Here *69*
Infrared Thermography *70*
Keeping Good Records *72*
Leak Detection *74*
Leaking Valves *75*
Lubrication *76*
Machines - Parts *77*
Detecting Mechanical Looseness *78*
Operating Vertical Equipment *80*
Remote Leak Detection *81*
Risk *83*
Run to Failure *84*
Shop Practices *85*
Spare Parts *86*
Stored Machines *87*
The Cost of Air *88*
The Little Things Matter *89*
Tunnel Boring Machines, Drill Head Bearing *90*
Unexpected Consequences *92*
How's the Weather? *93*
What Does That Warranty Cost? *94*
Your Brand New Machine Is Broken! *95*

Vibration Analysis 96

Basic Vibration Analysis *97*
Characteristics of an FFT Indicating Misalignment *99*
Costs of Resonance *101*
How Much is Too Much? *104*
Resonance *106*
There Are Limits *107*
Time Domain Spectrums *108*
Vibration Analysis Tools, gE *110*
Vibration Analysis: A Maintenance Information Tool *112*
What Causes My Machines To Vibrate? *113*

General Maintenance

Bearings

Analysis of Very Slow Bearings

Even after being in this business for 30 years every now and then someone will come along with a question that I have never heard or seen in person.

A mining engineer sent in the FFT spectrum below.

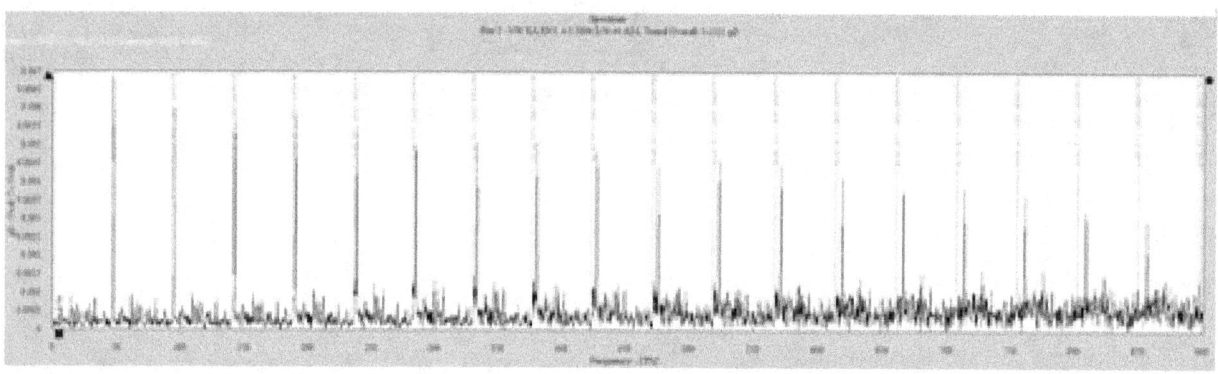

Although the FFT didn't transmit too well, at first glance it appears that there is damage to the installed SKF 23184E based on pre-calculated outer race frequency harmonics. But in fact there was nothing wrong with this bearing.

The SKF Bearing Handbook has the following note concerning slow speed bearings.

> When BPFO is being seen in slow rotating equipment, extreme care needs to be taken into account to avoid "removing a good bearing", because of possible "roller shuttle phenomenon"

The explanation is that at higher speeds centripetal forces keep the rollers in contact with the outer race. At slow speeds, rollers have more freedom to move around in the cage pockets so that rollers shift in the pocket as they roll over the highest part of the race. The shifting rollers impact the cage at the same frequency as an outer race defect. This phenomenon can be solved by studying the time waveform, because some of the individual impacts will occur at random times whereas damage impacts will have a consistent time between impacts. (assuming the RPM is consistent)

I think this would also be a good time to apply fresh grease as this would tend to reduce pocket movement, which could be seen in the time but outer race damage would maintain the same waveform.

Axial Loads On Rolling Element Bearings

Axial loads on a bearing that is not designed to carry axial loads will result in reduced bearing life and possible catastrophic failure. This is true no matter what the RPM of the machine. The subject motor bearing was driving a gearbox that over time had shifted from its designed installation position. This resulted in an axial load on the motor bearings, a force they were not designed to carry.

This picture shows clearly the results of an axial load on an NSK 6002 DU bearing with a C3 clearance.

This is a very graphic illustration of the damage caused by axial loads on a ball bearing.

Bearing Cage Failures

There are two aspects of a bearing cage failure, one is that they are fortunately rare and second, when they do occur, the damage is such that it is difficult to determine that a cage failure caused the problem.

Vibration monitoring with enveloped acceleration frequency analysis of the bearing components is the ideal way to monitor bearings. It is advised to trend the data and set the alarm limits so that an alert is given with an increase in the trend amplitudes. Even with due diligence cage degradation can occur very rapidly and maintenance may not be scheduled with enough lead-time to prevent a failure of the cage. For example, a customer was monitoring the cage frequency, noted that the trend of that frequency was increasing and notified maintenance. No action was taken and 23 days later the bearing failed.

Another customer provided the following frequency plot with the cage frequency (FTF) noted. In this case, the amplitude is well above the upper limits for replacement and they were fortunate the bearing was immediately changed. The two figures are the FFT and a picture of the bearing after removal.

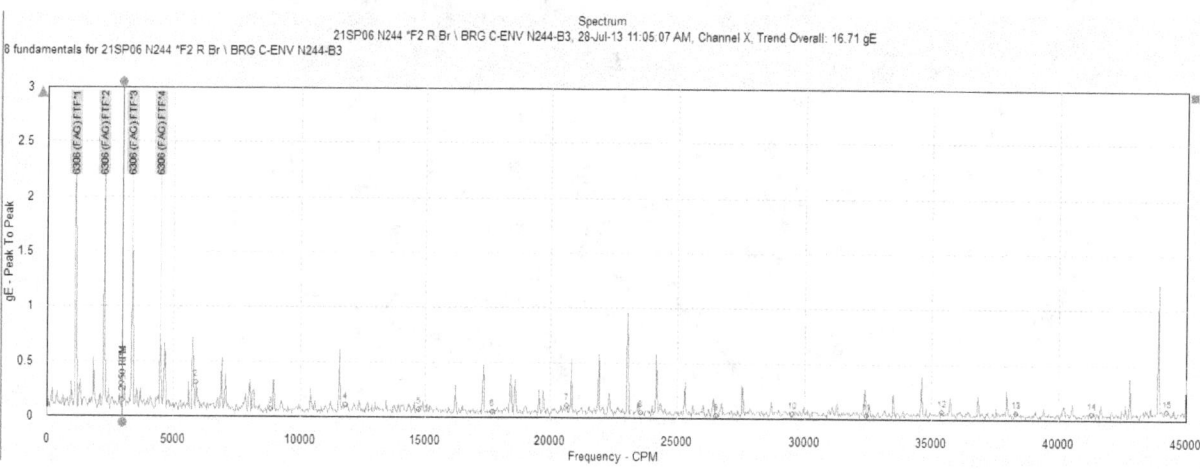

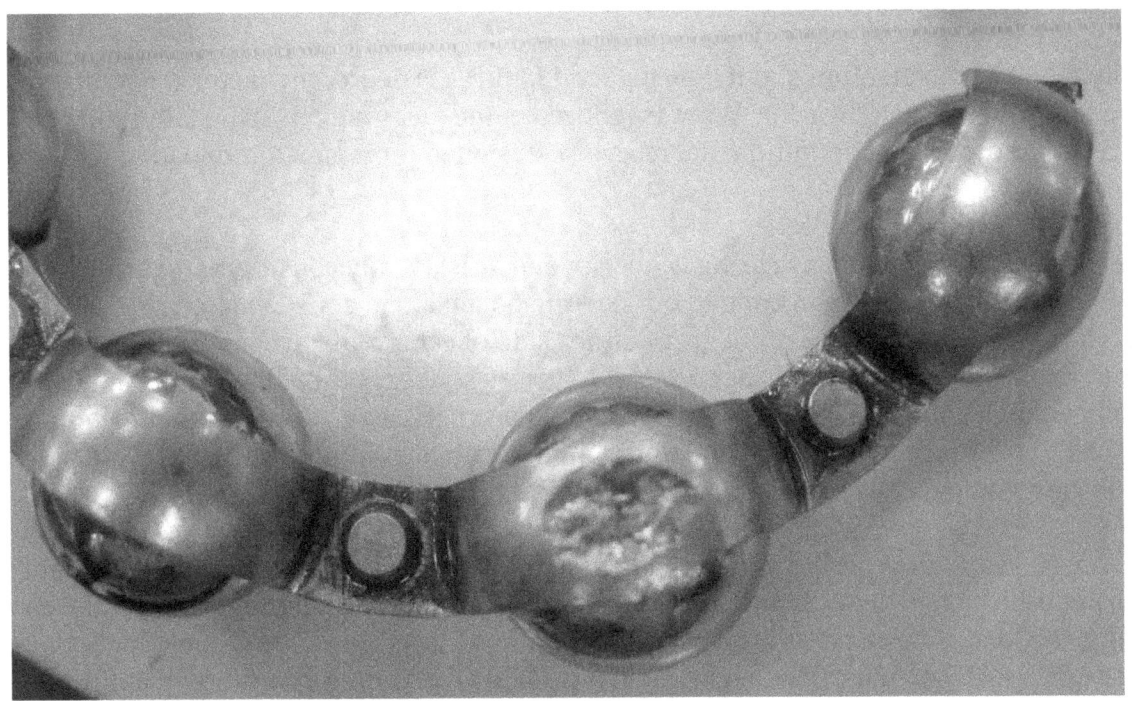

The lesson here is that it is not a good idea to wait after determining that the cage has developed a problem. In the second case, the amplitude is well above the recommended replacement value of 1.0 gE. It is best to be safe than sorry.

If you are having what appears to be an unusual number of bearing cage problems it would be advisable to review your bearing mounting procedures. If improper tools are used to mount the bearings, it is possible to damage the cage, which will later lead to failure. Mounting damage may not be visible with a casual inspection.

Bearing Clearances

Generally when a conversation about bearing load failures is heard, the factor most often discussed is an overloaded bearing. What is ignored is the fact that a bearing failure may be caused by an under loaded condition. Bearings are designed for a maximum load and for a minimum load.

Consider what happens when a bearing is overloaded. It is usually considered overloaded when the rolling elements are in contact outside the normal load zone, which varies with the type of bearing, but typically 10-15 degrees either side of center and are in metal-to-metal contact in the normal load zone. When excessive forces squeezes the lubricant from between the rolling elements and the ring, contact occurs and a small area of metal-to-metal contact occurs. The two surfaces are micro-welded together for an instant until the rolling element moves on, and you have the beginning of a fatigue failure.

The weld area is microscopic but will spread through the load zone as each rolling element rolls through the load zone. With the usual setup where the shaft is turning with a stationary outer ring, the load zone remains in the same location and is subject to this constant welding and breaking loose for each rolling element. There is no way to predict what will happen when this occurs except to know the bearing will fail, probably in the near future. Since the outer ring is being damaged, an FFT vibration signature will present a BPFO signal and it's possible that the damaged rolling element will be seen as a BSF signal.

An under loaded bearing is just what it says. Although somewhat rare, the load on the bearing is not sufficient to maintain the rolling elements in contact with the outer rings and after they pass through the load zone they begin to slide, rather than roll, until they again enter the load zone. Examination of the bearing will show shiny metal in the sliding area. The vibration signature of the FFT will indicate looseness and impacts as the rolling elements are trapped between the two rings and begin rolling again.

Whenever you have to use replacement bearings, from a different company, for example, check the bearing tables for proper clearances and insure that they are designed to operate under the loads generated by the machine. Each bearing manufacture publishes charts for their various bearings with the proper clearances. A bearing with improper clearances is either going to be over loaded or under loaded depending on whether the clearance is too low or too high.

Cage Materials

We are often asked about monitoring bearings and their various components and a variation concerns bearings cages. The cages are made of a number of materials, primarily brass and steel. As with the other components, rings and rolling elements, a damaged cage will generate a unique frequency based on the bearings size and speed of rotation. Although one can do the math to calculate the fault frequencies there are several computer programs that will quickly provide the information.

A current question was in regard to the difference in fault frequencies for an NU-218J bearing cage made of steel vs. an NU-218M with a brass cage. These designations are listed in the SKF Bearing Maintenance Handbook with other designations for machined, alloy, and plastic cages.

For an example, we obtained the fault frequencies for an NU218 and an NU-218E from the SKF Analysis software.

Where "X" is the rotation speed, the fault frequency for NU-218 is 0.428X and the frequency for NU-218 is 0.424X. Yes there is a difference but it is clearly insignificant. As noted, there are bearings with plastic type materials. These cages are so light that it is doubtful that an observable signal could ever be seen.

Although, except for the special cases, the cage material has little effect, variations in bearing sizes from one manufacturer to another can create major differences in fault frequencies. For example, some bearings with the same designation manufactured by two different companies will have a different number of rolling elements. Don't assume what is installed.

Cage Problems

It is all too common to observe at a construction site, new production equipment sitting at various locations covered with plastic or a tarp. They have arrived before the building was completed so are stored in the field. If this occurs over an extended period of time, the end results can be damaged bearings. No matter what time of the year, the bearings get warmer in the daytime and cooler at night, producing condensation. When this condensation occurs inside the bearing, trouble begins in two forms. First the hydrogen molecule in the water has a strong affinity to steel and attaches to the steel molecules resulting in hydrogen embrittlement. Second, the oxygen oxidizes the metal and you have rust. Then several months later when the equipment is installed and activated, loud grinding and scraping noises emit from the bearings. This was the case at a new plant in Richmond, Virginia. They were able to obtain seven of the needed eight replacement bearings from a local bearing shop but could not locate the eighth from any of the bearing shops in the area. In desperation they obtained a bearing from a used parts shop and proceeded with the installation. When this machine ran, it was much more noisy than the other three and we were called in to determine the cause.

First order of business was to collect an FFT as seen below.

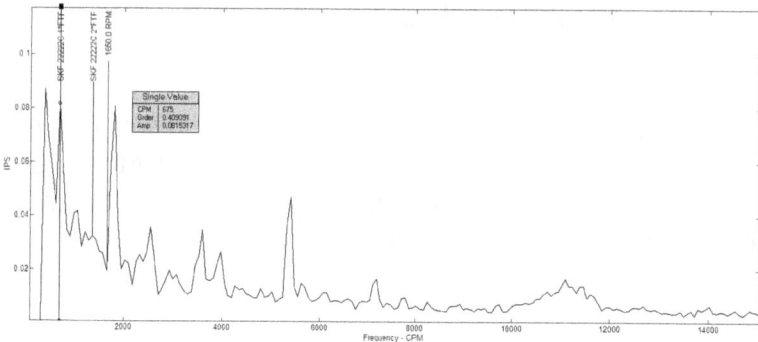

This overlay indicates for an SKF 22222C, a problem with the bearing cage (FTF). At this speed, the calculations for a cage fault computes to be 675 CPM as shown on the data box.

The bearing was removed and this photo shows the damage to this used bearing.

At some time in its previous life it had been either installed with a hammer and a drift pin. This damage was enough to distort the cage and result in an unsatisfactory performance.

The morale of story is not to install used bearings and when you install new bearings, use approved tools and methods.

"Checking" Bearings

There is probably not a mechanic or engineer who at one time or the other hasn't picked up a new deep groove ball bearing, stuck two or three fingers into the inner ring and gave the outer ring a spin. When asked what they are doing, a common answer is, "Oh, just checking the bearing." No, all you are doing is spinning the outer ring. The bearing has been tested after assembly at the bearing factory. There it was checked on automatic testing equipment where it is loaded and checked for precision of rotation, noise and smooth running. It is ready to use out of the box, all the user has to do is apply the proper lubrication and install it correctly.

What usually happens is that the person spinning the outer ring will complain that the bearing has too much clearance and he/she can hear a ticking noise as the outer race turns.

If a customer tilts the inner ring while holding the outer ring, a "big play" may be felt. The reason is that a deep grove ball bearing has an axial clearance, which is approximately 10 times the radial clearance. If the customer sends the bearing in for an inspection, which takes a special machine and set up, they are always found to be O.K.

There is a reason for the ticking noise. What occurs is that with today's packaging there is only a small amount of anti-corrosion protection injected into the bearings for shipping. When the person holds the unloaded bearing in the horizontal position and rotates the outer ring, the unloaded balls are "falling down" within the cage pocket clearance and this gives the ticking sound. In the vertical position all balls take a small contact angle and "roll" between the inner and outer raceway, hence no more falling within the cage pocket. If the user is not satisfied at this point, they can just add some clean lubricant which will dampen the ball movement and the ticking will stop.

How Bearings Are Analyzed Without Removal

As a rolling element travels around the bearing ring, if there is a damaged area, called a spall, it will create a vibration in the bearing where the frequency of the vibration can be determined by formulas detailed in previous blogs. If you run your thumbnail down a comb it will make a sound (vibration) where the frequency depends on the number of teeth and the speed of your thumb. Likewise, the speed of the bearing rotation, the bearing size, and the number of rolling elements determines the frequency of the generated vibration. In this case the number of rolling elements is the same as the number of teeth and the spall is the "thumbnail." Of course with a rolling bearing, the string of rolling elements can be considered infinite and the vibration is continuous.

In frequencies lower than 15-20,000 Hz, these vibrations can often be felt when you place your hand on the bearing cap. However humans cannot translate the vibrations into a specific frequency, we just know that it is "buzzing" and higher frequencies cannot be detected by humans. To accomplish the task of identifying frequencies we use an accelerometer.

An accelerometer is an instrument that contains a piezoelectric crystal, usually manmade, that has the characteristic of generating an electric current when it is squeezed. The crystal is entrapped inside the accelerometer housing between to brass slugs and connected to the outside by two wires. When placed on a vibrating surface, a bearing cap for example, the movement causes the two brass slugs to squeeze the crystal. The higher the vibration then the more it is squeezed and the higher the current generated. What makes it all work is the fact that the amount of current signal is proportional to the amplitude of the vibration so we know the amplitude. The frequency of the signal is mathematically determined by a process known as a Fast Fourier Transform or FFT.

Additional signal processing presents this information to the user on a portable or permanently mounted display. The following plot is from a very slow bearing rotating at 8.3 RPM and using enveloped acceleration to develop the data.

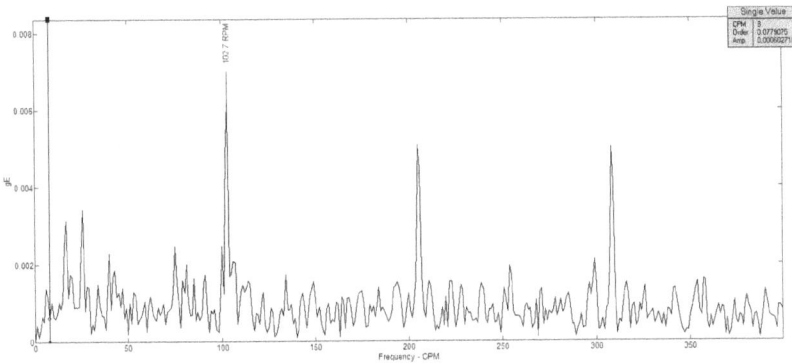

This would be considered a very slow bearing and requires the use of enveloped acceleration to detect the damage. The fundamental fault frequency, BPFO, is seen at 102.7 CPM, calculated by the embedded software. The amplitude is seen as 0.007 gE. A low amplitude but at this speed an indication of damage which is further seen by the presence of harmonics of the BPFO at 215 and 308 CPM

The bearing was removed and replaced before it failed. Maintenance under controlled conditions is always better than after a failure. Studies have shown that costs savings when maintenance is scheduled vs. "run to failure" are in the range of 10 times less.

Determining Damaged Bearings Without ID

Experience has shown that many times the owner of a machine has no idea what bearings are installed. Often the machine has been in service many years with several overhauls by different people and no one remembers what bearings were installed. In such cases, a helpful characteristic of the bearing fault frequency calculations is that when the contact angle is greater than "0", the multiplier will result in a frequency that is a non-integer multiple of the shaft speed. In Figure 1, the cursor is placed on an unknown frequency spike and the Order information in the Single Value box tells us it is 6.046X. We then place the harmonic marker on this mystery frequency and see that we have harmonics. Based on this information it would be prudent to do a physical inspection of the bearing as harmonics are an indicator of damage. The author has detected a number of damaged bearings using this method.

Figure 1 Unknown Bearing @ 6.04X

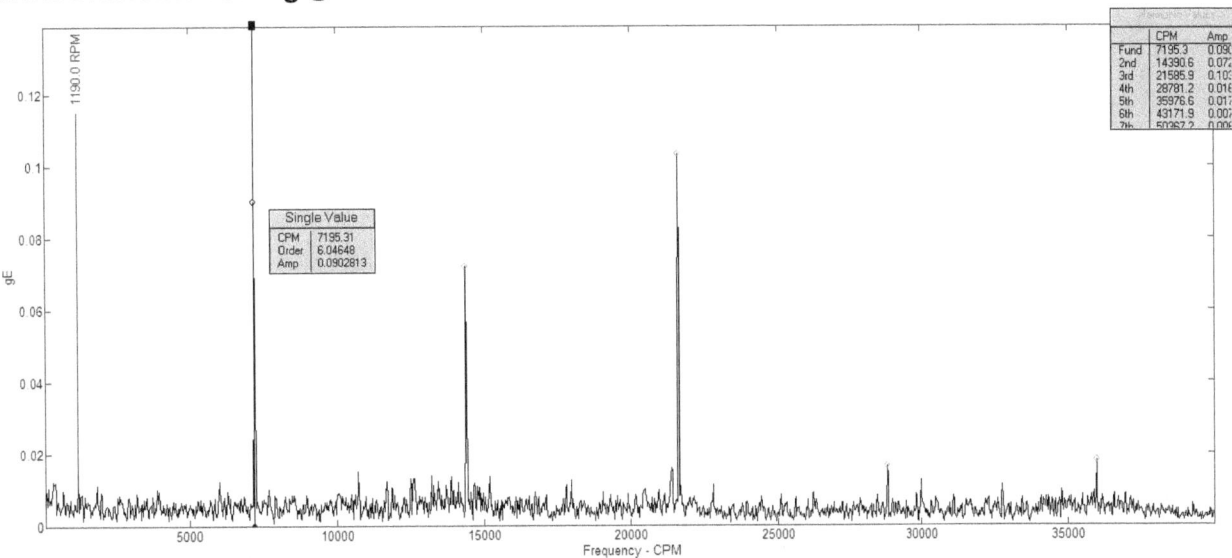

In Figure 1, a bearing had been damaged by the owner before our arrival. He wanted to see if we could find it among several others in the compressor. We did although the damage was only several small scratches in the outer ring.

Remember that the computer bearing fault frequencies are calculated based on new bearing dimensions. The bearings you are inspecting are possibly worn and consequently the actual frequencies generated may not fall exactly on the calculated frequency.

We have found through experience that approximately 90 percent of inner ring failures are caused by poor installation techniques. When the bearing is placed on the shaft by pushing on anything but the inner ring, with the proper tools, damage will occur. Force on the cage will damage the cage and push the rolling elements against the lip of the races causing damage to the rings. Even if the damage does not influence the machine operation, it will result in noisy bearings.
Bearing ball and rollers are harder than the rings, 61-62 Rockwell vs. 58-59 Rockwell. Therefore, damaged rollers and balls are rare during normal conditions.

Care should be taken to prevent water from mixing with the lubrication. 1 percent water in the lube system will reduce the life of the bearing by 90 percent. Water dissolved in the lubricant should not exceed 200 PPM

And finally, over 75% of machine failures are caused by the loss of the rolling element bearings. Why? Because of misalignment! Other than thrust bearings, rolling element bearings are designed to carry a radial load. When misalignment occurs, an axial force component is generated. When this becomes excessive, the bearings begin to fail. Probably the one procedure that would save the most money in any maintenance department would be to improve the alignment methods using laser alignment equipment.

It Bears Repeating

How many times have you been told to take care of new bearings? Treat them with the respect you would have for an expensive precision tool that is essential for the performance of your job. If you give it some thought, that's an accurate description of a bearing of any size. And I mention size because a national candy company decided it didn't need to monitor the bearing conditions in any motor less than 5 HP. Several years later they had a bearing failure in a 5 HP control drive motor. The failure of that motor caused the production line to shut down with product in process. I was told later that the total clean up and production loss was in excess of $250,000. All for a small motor bearing.

In South Texas they build oil-drilling platforms used in the Gulf of Mexico. In due time support equipment is added to the platform as it grows in size. At some point it was decided to install the anchor chain drives early in the construction. These machines are designed to keep tension on the anchor cables after the platform is in position. The timetable had installed the drive units over a year before the platform was completed. The bearings were 36" plain bearings and they sat out on the platform in the gulf's saltwater environment, uncovered, for that time period.

As the construction was completed someone decided that they should have the bearings inspected prior to commissioning the platform. Tests were performed that indicated that all was not well with the bearings condition. All four units were removed from the platform and dissembled in the shop. The damage was water etching on the surface of the bearing face. The damage was so slight you could only feel it by dragging a fingernail over the surface, but it was there and repairs were necessary. Later no one wanted to talk about the situation but the cost must have been very large for the repairs plus the downtime delay of the commissioning.

One common practice of construction companies is to buy the plant equipment at the start of construction and have it readily on-hand when it is time to be installed. This was done for four blowers at a site in coastal South Carolina. We all know that it is humid in South Carolina. During the day the blowers would be heated by the sun and in the evening would cool off and be covered by condensation. Apparently everyone thought that dew was harmless, thinking that it is just H20. Sitting there for some 9 months they were well saturated in all the nooks and crannies of a typical blower. And it seeped into the bearings. Hydrogen molecules attached themselves to the bearing rings resulting in hydrogen imbrittleness. Within 6 months of startup all eight of the 3.5" shaft bearings had to be replaced.

And finally there is the crew that takes the new bearings out of the box and leaves them on the workbench while they remove the old bearings and stir up the dirt and grit that goes along with that job. They wipe off the new bearings, install them and wonder 8 months later why they have to do the job all over again. Must be poor quality bearing they surmise.

The moral of the story is to treat your bearings with care; from the day they are purchased until the day they are installed and back on-line. Remember, you are working with a precision instrument.

Monitoring Ultra Slow Rolling Element Bearings

Very slow rotation occurs in such units as cranes, radar antennas, steel mill ladles, anchor chain reels and radio telescopes. For example the Very Large Array (VLA) consists of 27 radio antennas in a Y-shaped configuration on the Plains of San Agustin fifty miles west of Socorro, New Mexico. Each antenna is 25 meters (82 feet) in diameter, weighting 230 tons, resting on approximately 10-foot diameter bearings with 156, 2-inch rollers. The data from the antennas are combined electronically to give the resolution of an antenna 36km (22 miles) across, with the sensitivity of a dish 130 meters (422 feet) in diameter.

How slow is ultra slow? These bearings rotate at a rate of one revolution in 9 minutes on high speed! During operation they match the speed of the earth, once per 24 hours. Although using enveloped acceleration technology to monitor a bearing rotating at 1800 RPM provides excellent data for a highly accurate FFT analysis, when bearings rotate this slow they do not generate the frequencies needed for a standard vibration analysis.

After a catastrophic failure of one bearing, the operators were searching for another way to monitor the telescope bearings. Based on grease analysis, several units were suspect but because of the high cost and complexity of replacing a bearing, and with only one spare, it was highly desirable to replace the one most degraded of the 27 bearings in service.

From the physics lab we recall: ***F = MA*** where **F** = Force, **M** = Mass and **A** = Acceleration. Since the mass of each unit is the same, we can set it equal to "1", then the amplitude of the acceleration signal is proportional to the amount of force generated inside the bearing. This in turn is a measure of the amount of damage the rolling elements were passing over because it requires more force to move the rollers through a damaged area. When the time domain spectrums were collected and ranked by overall amplitude, Unit 23 was found to have an amplitude more than 3 times greater than the second highest. (0.34g's vs. 0.1 g's) That was the bearing that was changed.

Having more than one similar bearing helps in such evaluations. If only one of a kind is available then it is necessary to collect data over a period of time, 6 months usually, and plot a trend of the amplitude. An increasing trend is a warning sign and a tear down with a physical examination may be necessary.

Overheated Bearings

I was recently asked what were the more common reasons for a bearing to overheat. This lead to a research of various published trouble shooting guides and this list was composed. As with any list, it is probably not complete, there is always something else but here is what we have with some comments. There are special cases but in general, motors, fans and pumps should maintain a temperature below 180 degrees Fahrenheit. Extreme temperatures will alter the grain structure of the bearing material leading to failure

1. Lack of lubricant. This is rather obvious, without proper lubrication, metal-to-metal contact occurs generating heat and destroying the bearing.

2. Too much lubricant. It has been said that more bearings are destroyed by too much grease than from not enough. Bearings are not designed to churn grease, which is what happens when the bearing is over greased. The grease absorbs this energy, heats up and it turn overheats the bearing. In addition, excessive grease can distort lightweight cages leading to displaced elements and undesigned loading.

3. Misfit of bearing in housing. This can go two ways, either too tight or too loose. If it is too tight then the lubricant is squeezed out of the rolling element/race contact point and metal-to-metal contact occurs. Then micro welding of the two components occurs and immediately breaks apart leaving a pit in the race. Over time this develops into a spall and the bearing has failed. For a bearing that is not properly loaded, the rolling elements slide outside the load zone instead of rolling. This also leads to mechanical damage and bearing failure.

4. The wrong lubricant is used. Bearings are designed for specific uses and likewise the lubricant is designed for specific situations. This subject is the basis for full college courses so needless to say, make sure the proper lubricant is installed. All of the lubricant companies publish charts for example, with regard to speed, size, environment, environment, temperature, and frequency of application and on and on.

5. The easy one, a grease plug is not secured or a seal fails and the lubricant leaks out and we have an overheated failure.

6. Then we get into the external forces that can override any of the above. A properly installed bearing with the proper lubricant can be put in a bind by, for example, misalignment. This can overload the bearing to the extent that clearances are removed or increased causing the conditions as covered in #3. The lack of shields can allow the product being manufactured to overheat the bearing.

Vibration itself doesn't cause vibration but the results of excessive vibration can lead to the conditions of #3. In some cases several causes work together with vibration to overheat the bearing.

In sum, one must install the bearing properly, lubricate it properly with the correct lubricant and operate within the engineering limits provided by the manufacture. And then after all that someone will come along with a fork lift, hit the motor housing, push it way out of alignment and cause a bearing failure because the operator didn't tell anybody what he did, nobody did a routine check and the motor ran to failure.

Analysis of Rolling Element Bearings

Since one of the analysis techniques involves trending of vibration levels, it is important that the data collection location be marked and the same location be consistently used each time data is collected. Common to most modern portable electronic data collectors is the accelerometer. These are generally constructed with a man-made piezoelectric crystal, which generates an output voltage directly proportional to the amount force applied. The accelerometer is usually placed on the bearing cap, or if not accessible, as near as possible to the bearing.

In those instances where it is not possible to safely position the accelerometer by hand, the accelerometer may be permanently stud mounted or glued to the machine and the signal wire terminated in a safe location. Generally the accelerometer will be mounted using a magnet. Both methods are acceptable for general vibration monitoring. In rare instances a stinger (a steel rod threaded onto the end of the accelerometer, do NOT use an aluminum rod) may be attached to the accelerometer to reach a bearing cap located in a tight space, but stingers will alter the signal amplitude and frequency and are not recommended for general usage.

For continuous monitoring of a machine, all of the points of interest use a stud or epoxy mounted accelerometer. The signal wires are then terminated at a common point where they are multiplexed and routed to a permanently mounted data collector. The signals from the data collector are then passed to a computer controller, which is programmed to store and process the data. One accelerometer signal can be processed into four presentations; acceleration, velocity, displacement, and enveloped acceleration and these may be processed for different frequency ranges as needed. In other words, the acceleration signal may be presented in one spectrum from 0-30 Hz in velocity to check for balance and alignment. A second spectrum may be generated with a range of 0-1000 Hz to disclose the rotor bar pass frequency. The acceleration signal can also be processed with enveloped acceleration algorisms to check for bearing degradation and finally a high frequency 0-10,000 Hz acceleration signal may be processed to provide a gearbox inspection. In addition other types of sensors can collect operational data such as shaft position, speed, temperature, flow, pressure, etc. Generally any sensor that provides a voltage output can be monitored; the signal can then be collected and stored for evaluation.

Rolling Element Bearings II

A rolling element bearing consists of four components. The parts and their fault frequency abbreviations are: the outer ring, Ball Pass Frequency Outer, (BPFO), the inner ring, Ball Pass Frequency Inner, (BPFI) the cage, Fundamental Train Frequency, (FTF), and the rolling elements, Ball Spin Frequency, (BSF). Each of these components will generate a unique frequency, which will vary in relation to the shaft speed. Therefore an accurate rotation speed is required for the analysis. Over time the bearing parts will wear and the observed frequency may vary slightly from the calculated frequency and manufacturing variations are a factor. As can be seen in the following frequency calculation formulas, the frequency generated is based on the number of rolling elements, the shaft rotation speed, ball diameter, pitch diameter, and the contact angle. These are provided although all of the data collection software on the market will do the calculations if the user provides the bearing ID i.e., an SKF 6309 or FAG 22222.

Bearing frequency formulas
(1) $BPFO = (N/2)(RPM/60)(1-Bd/Pd)(\cos\phi)$
(2) $BPFI = (N/2)(RPM/60)(1+BD/Pd)(\cos\phi)$
(3) $BSF = (N/2)(RPM/60)(1-[BD/PD]^2)(\cos\phi)$
(4) $FTF = (1/2)(RPM/60)(1-Bd/Pd)(\cos\phi)$
And:

N = Number of balls or rollers
Bd = Ball diameter (in or mm)
Pd = Bearing Pitch diameter (in or mm)
ϕ = Contact angle, ball to race

These formulas are for the bearing mounted on the shaft and a rotating inner ring. If the outer ring is rotating, reverse the (+) and (-) in the formulas. Another handy rule of thumb to use when you are in the field and the computer is back at the office and you need a close approximation:

BPFO = (RPM) (n) (0.4)
BPFI = (RPM) (n) (0.6)

The first four formulas will give the frequency results in Hertz (Hz). Hz is cycles per second. If you desire them in cycles per minute, (CPM), multiply by 60.

Vibration amplitudes are measured in the following units:

Displacement (distance) is measured in Mils, one mil equals 0.001 inches. Metric measurements are in millimeters.

Velocity (speed) is measured in Inches Per Second, IPS. For metrics, the units are mm/sec. For a quick approximation, 1 mm/sec equals 0.04 IPS. 25.4 mm/sec equals 1 IPS.

Acceleration (force) is measured in G's, for both English and Metric units. An acceleration signal may be integrated to produce a velocity signal and double integrated to produce a displacement signal. Most software programs will do the computations.

Enveloped Acceleration is a special measurement of acceleration, gE, and there is **NO** comparison or conversion to the standard measurement methods.

Sealed or Open Bearing?

It is well known that some folks learn the hard way.

After his description of the operating conditions of the motor, we recommended using a sealed bearing. The description included rain, dust and heat. After several exchanges of correspondence where the customer kept asking why he couldn't use an open bearing and our explanation of why it probably wouldn't work, he said he was going to use an open bearing.

To his credit, several months later he wrote back and sent a picture of what the bearing looked like after a short time in operation. One wonders why he asked in the first place if he was going to do what he wanted to anyway.

The bearing is rusted, the grease is discolored and the bearing destroyed.

The lesson, install bearings designed for the service environment where they will operate.

Shiny New Bearings

In the past 30 years that I have been in the plant maintenance business one topic that continues to come up for discussion, as it needs to be, is the proper handling and storage of bearings. A real life example clearly makes the point.

A well-known manufacturer of "quiet" motors filed a complaint about the noisy bearings they had received. The motor they manufactured was predominately used in dentist offices where they prepared dental appliances. Their complaint was that the motors made a disturbing clicking noise when operating.

We went to the manufacturing plant and observed their procedures. There was not a continuous demand for these motors so they would periodically run a batch and place them in storage. In the same manner they would order the component parts in batches and place them in storage.

Then, proving the point that no matter how many times you tell someone how to do something, they are going to do what they want to do when the time comes. Disregarding the advice that when you get new bearings you leave them in their wrappings and in the box, the storeroom clerk decided that the shiny new bearings looked better unwrapped and placed, laying flat, on the shelf. Adding to the problem, the storage environment was not environmentally controlled. Located in a humid industrial area, down by the river, the air was slightly acidic. A cold bearing, sitting open on the shelf, will attract condensate moisture as it warms from nighttime to daytime. A teardown on one bearing revealed that the acidic moisture had corroded a small spot on the down side of the bearing and on the bottom of the ball. When the bearing was installed in a vertical position, the damage on the sidewall of the ring did not make contact with the ball but the ball made normal contact with the bottom of the ring. Since a spot on a ball will only make contact with the ring about 11% of the time, the damaged balls generated a rather random clicking noise although the identifying frequency was the BSF, Ball Spin Frequency.

To say it again, don't unwrap your bearings until the moment you are ready to install them. Yes, they are pretty and shiny but they are industrial equipment, not works of art for display.

Shop Practices: Handling Bearings

Not many of us have the opportunity to work each day in clean rooms like we see in articles about satellites and medical research. Most of us do not even work in a shop that is air conditioned unless you consider the open windows and bay doors as providing "air conditioning". Consequently we have to always be on alert about the environment when we are working with rolling element bearings.

This applies from the day the bearings are received into the storeroom until the bearing is installed and in service. Prior to shipment, the bearings are coated with a preservative and sealed in a coated paper then sealed in a box. The manufacturer insures that the bearing is clean and ready for installation. But here are some of the shop practices that have been observed many times that will reduce the bearings life.

1. After withdrawal from the storeroom, the bearing is unwrapped and placed on the worktable near where an old motor is being blown down with a high-pressure air hose.

2. While installing, the bearing is dropped and bounces on the floor, but looks "O.K.".

3. If 3oz of grease is good, then 5oz forced into the bearing and housing is better.

4. You don't have a mounting tool but use a wooden hammer and just tap it gently around the ring until it is seated.

5. Lunch time? Just leave the bearings on the cart; we'll finish up when we get back.

6. Alignment is no big deal, just use a straight edge and line up the coupling halves.

Airborne particulates, depending on their size, will embed in the rings and be the beginning of a spall, which will lead to failure. Anytime a bearing is dropped, it should be discarded. Point contact from the rolling elements to the rings will start a false brenelling spot on the ring which depending on the location, can grow into a spall. The recommended grease schedule is designed for maximum life. Bearings are not designed to plow through grease and the churning will generate unwanted heat. Tapping on a bearing with any type hammer is foolish. Again false brenelling can occur, reducing the life of the bearing. That bearing lying on the cart is open to all the particulates in the air. And finally, how many people can align two parts to within 0.003" with just eyeballs and a straight edge. Misalignment is the most common cause of bearing failures.

Look around your shop, if this type of situation is present, make some changes and improve your bearings service life.

Unusual Characteristics of Very Slow Bearings

There have been a number of questions concerning the monitoring of slow speed bearings and the following is extracted from SKF Bearing publications.

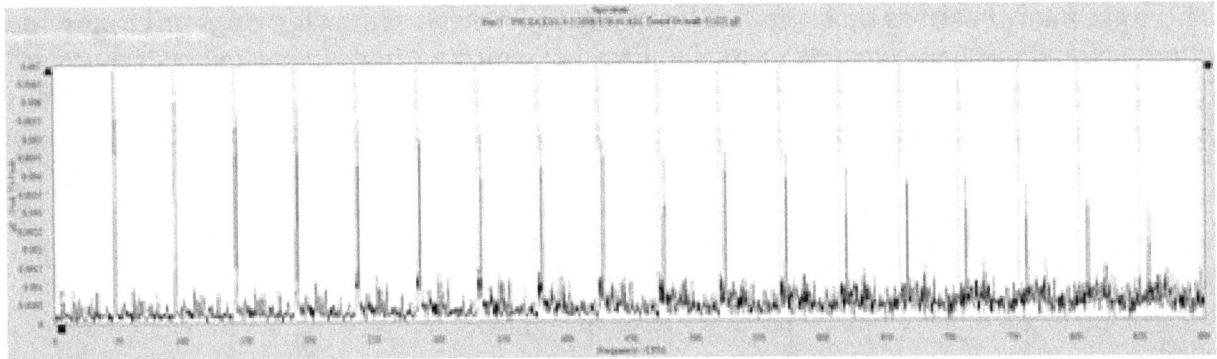

"The frequencies represented in the figure match the pre-calculated outer race frequency of an SKF 23184E.

When BPFO is being seen in slow rotating equipment, extreme care needs to be taken into account to avoid 'removing a good bearing', because of possible 'roller shuttle phenomenon' that shown on BPFO also!" **SKF**

Because the rollers are impacting the cage, the BPFO frequency is generated. The way to distinguish between this random occurrence, as the rollers pass over the top of the bearing, is to use the time domain spectrum. If there is actual damage in the outer ring, the impacts will be consistently at the same interval. If the signals are generated by the rollers "rattling around" in the cage, they will generate a random time interval.

Very Slow Bearing Analysis

One question that is continuously asked on the SKF Reliability Forum concerns monitoring very slow bearings. The short answer is that it is possible and here is one example.

There are many bearings that rotate at less than 1 RPM such as slewing rings on cranes, radar antennas, and discharge vessels in steel mills, all of which we have inspected over the years. The following example is a bearing that was rotating at approximately 0.5 RPM. Operations will tell you that it is never convenient to shutdown for maintenance and in this case, loss of production plus replacement and labor would incur a cost of over $250,000. Very slow bearings require the use of the time domain spectrums. As can be seen in Figure 1, the FFT, where the BPFO is noted at 6.9 CPM, you are not able to distinguish the unique, very low amplitude frequencies and operational noise masks any other useful signals.

Figure 1.

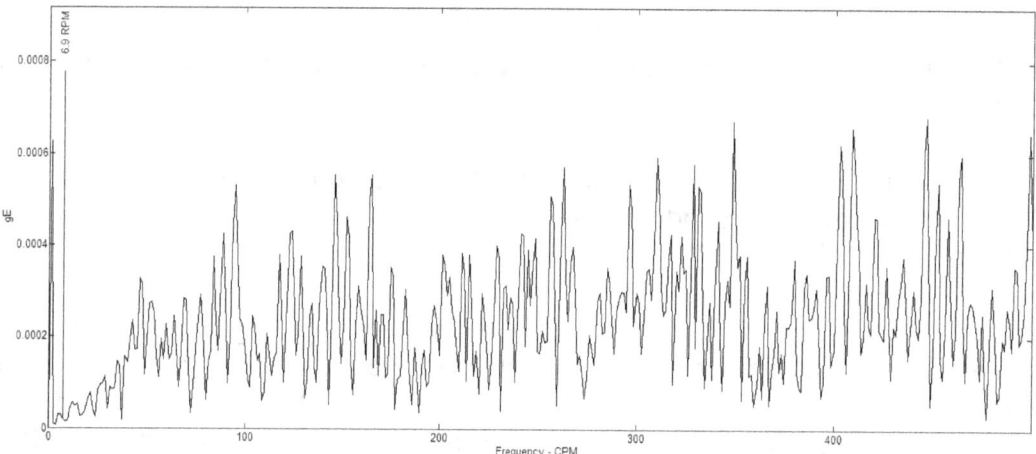

The solution is to use the time domain spectrum of the suspect bearing and compare it with a similar machine. You don't always have another machine to compare but when you do, you use it to reinforce your investigation.

Figure 2.

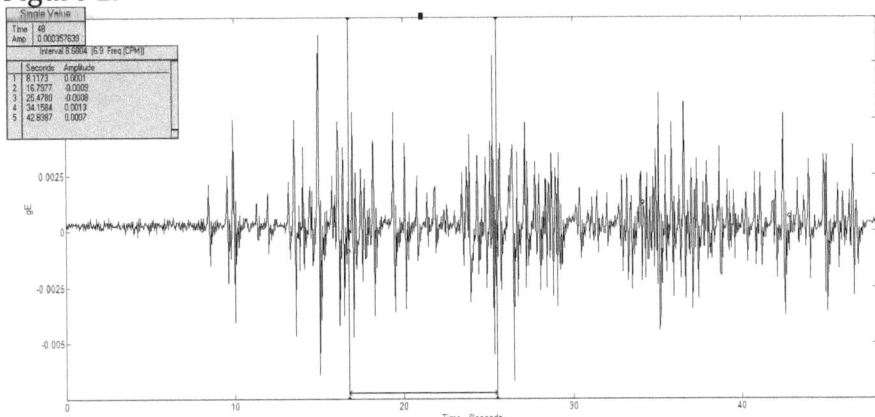

The BPFO of the bearing was calculated to be 6.9 CPM. The harmonic marker is used to set the left and right cursors at 6.9 CPM. Then clicking on the bottom of the enclosed field, then both cursors can be dragged to a position where they appear to be in the middle of the two pulses and will also line up with the next pulse. This portion of the analysis is a judgment call on what lines up and what doesn't. Note that the 8.6 sec interval is equal to a frequency of 6.9 CPM. A number of spectrums had to be collected to acquire this spectrum because it was found later that the bearing housing was being distorted so that the rolling elements were not always in contact with the ring.

For a comparison, the second machine was examined and resulted in the following FFT and time domain. Figure 3 & 4.

Figure 3

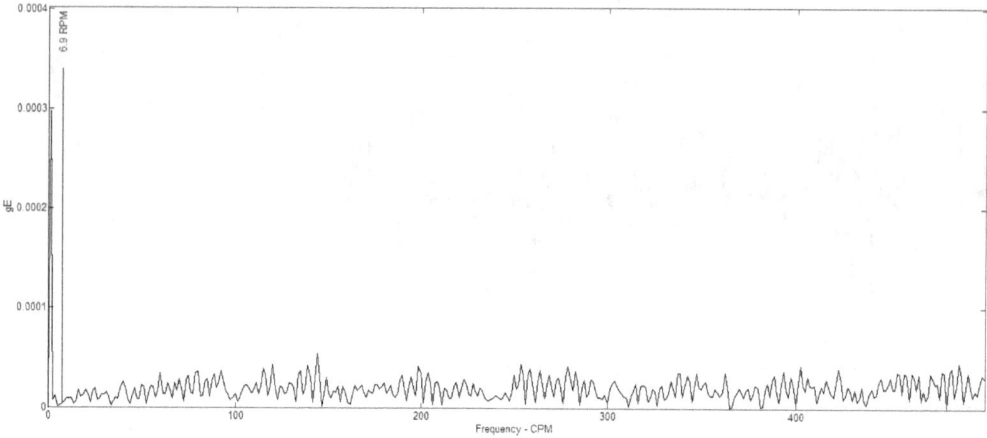

Figure 4

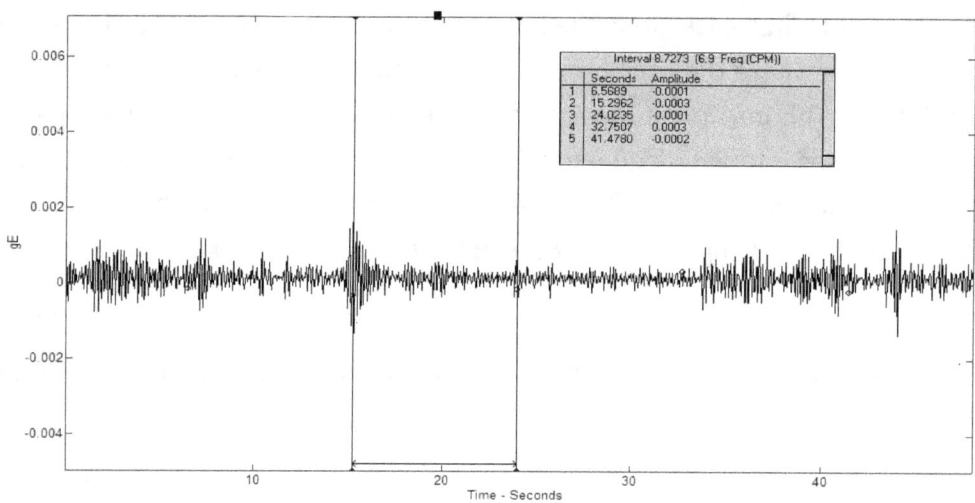

At the presentation for the manager, the two time spectrums were overlaid as seen with two different colors. With one shown in red and the other in blue and based on the overall amplitudes of the two spectrums, it was deduced that there was nine times more energy being generated inside the suspect bearing as in its twin. The decision was made to change the bearing and severe damage was found in the outer ring. It was also found that the mechanism designed to allow for heat expansion had failed and was the source of the housing distortion.

What's the Difference

We recently had a customer who was doing some home repairs write in to question the difference between two bearings that have the same measurements, 6.5mm housing for a 2.54mm shaft. Both are SKF bearings, SKF 613912 and SKF 1838001/c4. Then when he examined the bearings he said that he found that one had 9 balls and the other had 8 balls. He clearly had a problem because the SKF 183001/c4 has 15 balls. The difference is that he wanted a c4 clearance and he had probably purchased an 1838001 without clearance specifications. We weren't sure what his problem was other than he had the wrong bearing.

We did some research and found that the difference relates to the use of the bearings. In one case it is designed for a Vespa motor clutch (1838001) and the other is designed for the motor (613912). It is not at all unusual to have numerous bearings of the same size but what is not apparent is their application.

The lesson from this is to not buy bearings by their size but by what they are designed to do and this includes shaft rotation speed, loads, and type of lubrication, etc. For this you need to consult the technical parameters of the machine as determined by the manufacturer and follow their recommendations.

In this case the customer was about to put the clutch bearing in the motor, which because of the different load capacity, would probably had an early failure. The SKF dealer exchanged the bearing and that was the last we heard from him.

Although the 1838001 shown is not an SKF bearing, the difference construction of a c4 bearing is readily apparent in the pictures.

NBC Ball Bearing 1838001
Model: 1838001
Brand: NBC

SKF Ball Bearings 1838001/C3
Model: 1838001/C3
Brand: SKF

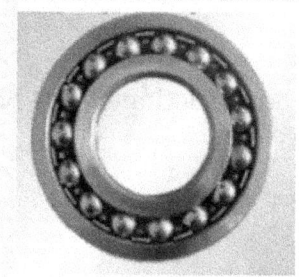

SKF Ball Bearings 1838001/C4
Model: 1838001/C4
Brand: SKF

Power

The title says "Power" and the specific power is electrical. Not too many years ago one of the essential tools that all maintenance technicians carried was a 50 or 100-foot electrical extension cord. Power for tools came from a wire and you had to have one. Today with the advancement in battery technology what most of us have is a total reliance on batteries to operate our tools. So instead of carrying extension cords, we carry a couple extra, fully charged batteries.

This was the case while we were on a job on an early spring day in Louisiana. Then we received a frantic phone call asking that we get over to Houston as soon as possible, there were some bearings on a drilling rig under construction that had to be checked before the next phase of the job could continue. So we packed up and hit the road. Arriving late the decision was made to leave everything in the trunk of the car, it would be safe until in the morning.

The next morning we proceeded to the job site and set up our equipment. The sleeve bearings of concern were installed on an anchor chain motor drive set. The 4 units had been installed over a year ago and had been stationary and unprotected. The prime concern was there might be water etching of the bearing surfaces and we would be looking for the additional roughness this etching would create.

All went well for about the first 15 minutes and our equipment began to fade out. The batteries were showing a complete discharge. But, not to worry, we always carry spares. These were installed and testing resumed for about another 15 minutes and then the spare batteries also showed a complete discharge. We were baffled with the situation when the second set of spares also faded after just a few minutes. We had a charging unit so in order to complete the testing, we cycled the batteries from the data collectors to the charging unit and swapped them out with just partial charges. It was embarrassing and did create some delays of course.

The question then was what happened to our batteries that caused them to discharge so rapidly. We reviewed what had happened and then recalled how cold it had been for a March day in Houston. Bingo!

All the equipment, data collectors and spares were stored in the car overnight and the temperature was in the 30's in the morning. A cold environment is a big deterrent to maintaining a battery's charge. The batteries had a full charge the day before and were operating normally but the unexpected overnight cold had sapped their strength.

A word to the wise then is to treat your battery operated equipment and their spare batteries with care and keep them comfortably warm during the winter/spring months.

General Maintenance

Electrical

Predictive Maintenance

One occurring theme on many maintenance forums is a question framed around, "What is Predictive Maintenance?" This is then followed by a purported serious discussion of the definition of the word "predictive" with the goal of being able to predict when a machine or one of its components will fail. Therefore, the question really becomes, "Can you predict when a machine or a component will fail?" The answer is: **NO**.

The author has been in the machinery diagnostics business for 32 years. During that time I have used and manipulated many software programs that were supposed to produce as their end product, a predictive time of failure or a time of reduced productivity when a machine should be removed from service. Without fail, none of them worked as advertised. The best any of them would do is to produce a series of "if" statements. "If" this is occurring, do this or "if" this is happening, do this, and so on. In real life, there are too many variables to incorporate into an algorithm that can produce a specific short-range date.

What can be done and is done on a daily basis all over the world is to develop machinery trends. One may <u>trend</u> oil particle counts, temperatures, vibration amplitudes or frequencies, pressures, cycle times, etc. All these parameters give us an indication of the health of the machine and with the accumulation of enough data, years, not months; the deterioration of the machine can be monitored. Then, based on experience one can compare what happened in the past to the current conditions and make an educated guess where conditions will be in the future.

For example if in the past when the pump pressure dropped by 5 psi the seals were found to be worn, then it's probable that if your pressure has dropped by 5 psi over the past 6 months you will find some worn seals. Or if a bearing fault frequency has slowly increased in amplitude over the past year and now exceeds a recommended amplitude for continuing operations, then you probably have a damaged bearing.

In any case, the technician should never provide a specific date for failure, that is a lose-lose proposition. If you state a date and on that date it is still operating, you lose. However if you state a date and it fails before that date, you lose. Always state any prediction as a date range based on current trends compared to past experience.

Variable Frequency Drives (VFD) Part I

Often when it is desired to have a variable speed fan, conveyor belt or mixer, for example, the solution offered is a Variable Frequency Drive (VFD). As the title states, the electronics are designed to allow the usual 50 or 60 Hz electrical power frequency provided to an AC motor to be increased or decreased in order to change the rotational speed (RPM) of the driven unit. This seems to be a simple and fairly inexpensive solution to acquire the ability to change the speed as needed by operations.

Two problems often overlooked are that by reducing the RPM of the unit the speed may move into a resonance frequency and the electrical interference generated by the electronic circuits of the 6 silicon rectifiers used to modify the driving electrical frequency.

Every mechanical device has a resonance frequency. This frequency is determined by the mass and stiffness of the unit. When a device is designed, the resonance frequency is mathematically determined. As a general rule this initial resonance frequency is designed to be at least 15% above and below the design RPM. For example, consider a typical machine designed to operate at 1,000 RPM. It could be safely operated in a range of 850 to 1150 RPM (1000 plus or minus 150). After a machine is constructed the response frequency is often determined by performing a "bump" test which simply stated is to impact the machine with a hammer and measure the frequency at which it "rings".

The problems begin to occur when it is determined that 850 RPM isn't what is needed and they need a slower or faster speed. A simple turn of a dial further reduces or increases the voltage frequency and the desired speed of 740 RPM or 1200 is produced and everyone is pleased. However this is outside the designed frequency speed range (850-1150) and just happens to be close to a natural resonance frequency.

Sometimes the unit immediately begins to shake or as often happens, the machine slows down or speeds up and everyone is happy. What is not seen is that a low or high amplitude resonance vibration is being generated which over an extended period of time will cause damage to various components. It might be the bearings; it might be some brackets or the foundation. The bottom line is that the machines components are slowly being beaten to death. A mental picture for you is to think of a tiny man with a tiny hammer and he is beating on the components 740 times a minute. The question is how long will it last? Then the answer is, how many times can you bend a piece of metal or plastic before it breaks. The answer is of course indeterminate, but sooner or later it will break. And in the worse case it's not a little hammer but a big hammer, and in short order the machine is out of service.

Once the natural frequency is determined, the control dial must be annotated with an allowable speed range and any dangerous speeds noted and avoided.

Part II of this blog will discuss the electrical problems generated by VFD's.

General Maintenance

Gears

Gear Analysis

It is not unusual for a gearbox to have more than one set of gears meshing while the machine is in operation. When doing an analysis of the gearbox, the initial task is to identify the meshing gears and their gear mesh ratios. Then starting with the input RPM, if the first mesh is a speed reducer to 1/2 the input RPM, then the gear mesh frequency of this set is (RPM)(number of teeth on the input shaft) i.e., (1000 RPM) X (24 teeth) or 24,000 CPM. Using the output from this set, 500 RPM, the same process is applied to the next gear set. When the output shaft RPM is determined, you have a setup of all the gear mesh frequencies that you can expect to see in the FFT.

Foreign material such as sand and grit are very damaging to gear sets. Silicon is harder than the metal in most gear teeth and the results is pits in the teeth as the teeth engage and press the sand into the metal. Forces in excess of 250,000 PSI are not unusual between the teeth of a gear set. These sand indentions then lead to spalling and pitting and a shortened life of the gear set.

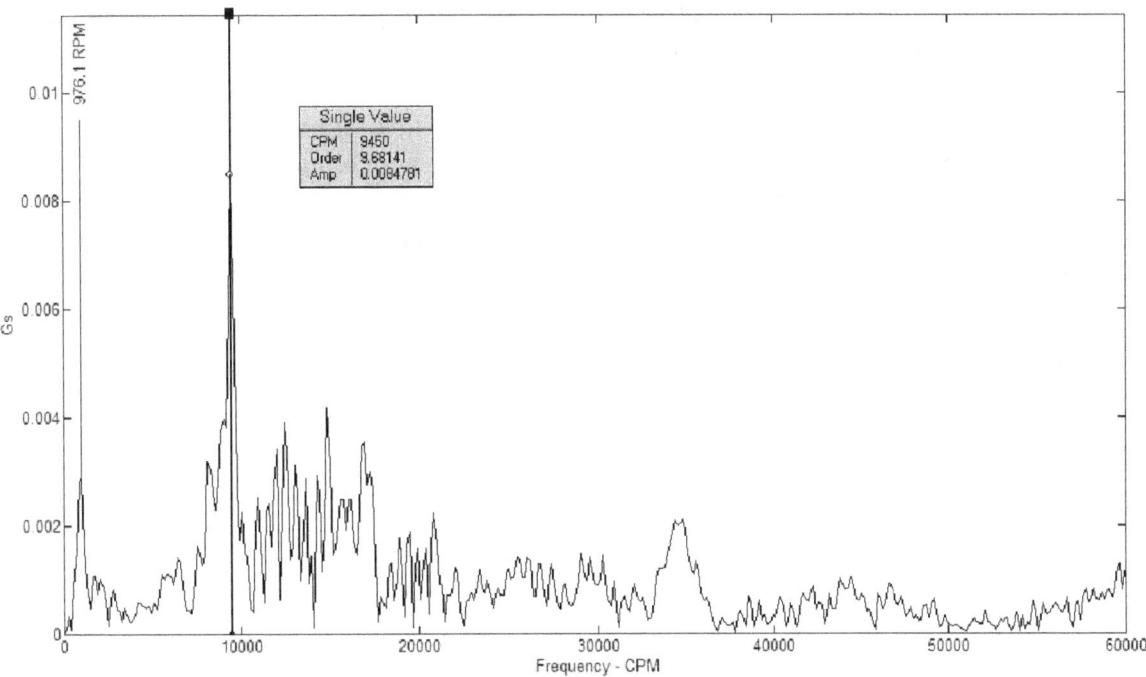

In this spectrum the gear mesh is located at 9,450 CPM and although the amplitude is low in this case, further damage will increase the amplitude over time.

Vibration analysis points out a problem with the gears that is developing; the root cause is contaminated lubricant. It would be wise to do a careful search to determine how the foreign material was inducted into the lubricant. If allowed to continue the next oil examination will probably discover metal particles in the oil sump.

Gearbox Maintenance

The case where a rather complex machine that can be easy to monitor and something almost every manufacturing plant has installed, is a gearbox. A gearbox can have several speed reductions or increase gear setups. In simple terms in a speed reduction, a 25-tooth gear rotating at 500 RPM is driving a 50-tooth gear, a 2:1 ratio. The small gear rotates 2 times for 1 rotation of the large gear so that the output shaft turns at 250 RPM. That shaft in turn may have installed on it another 50-tooth gear meshing with a 100-tooth gear so that its output is 125 RPM.

Just as in bearing analysis where the rolling element passing over a damaged area generates a spike of energy, as each tooth in the gear engages and drives the second gear, an acceleration force is generated. The frequency of this meshing of gears is equal to the (shaft speed) times (number of teeth) of the gear installed on that shaft. Example: 25 teeth times 500 RPM will equal a frequency of 12,500 CPM and the amplitude will be measured in "G's". Because there are always teeth engaging and driving the other gear, the gear mesh (GM) is always present on in the acceleration FFT. In the case of multiple gears installed, there will be multiple GM frequencies, one for each gear set. And of course, the other gear in this example generates the same frequency, 50 teeth times 250 RPM equals 12,500 CPM.

If the GM frequency suddenly increases and you hear a clicking noise emitting from the gearbox, you suspect that a gear has been damaged. But which one? Do you have a spare gear? Again, which one do you need a spare for? To answer this question of which gear is damaged we can look at the following FFT from a damaged slow speed gearbox with an input of 225 RPM.

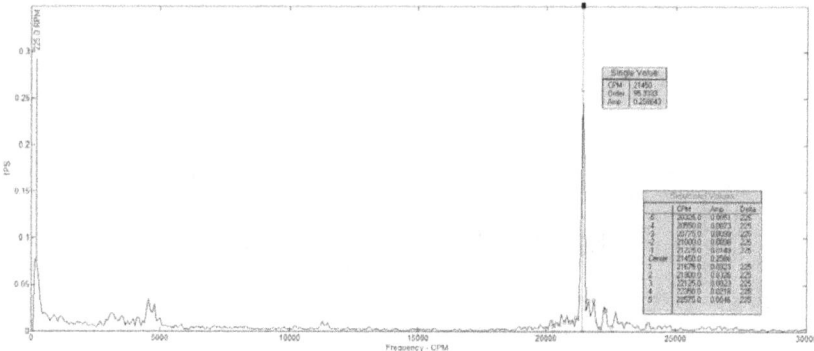

The spike of energy is the GM for this box, 21,450 CPM. Also included are markers on the sidebands that surround the GM frequency. The spacing of the sidebands is shown in the Sideband Values chart to be 225 RPM. This matches the input RPM so we know that the damaged gear is on the input shaft.

We also use the time domain spectrum to verify the information.

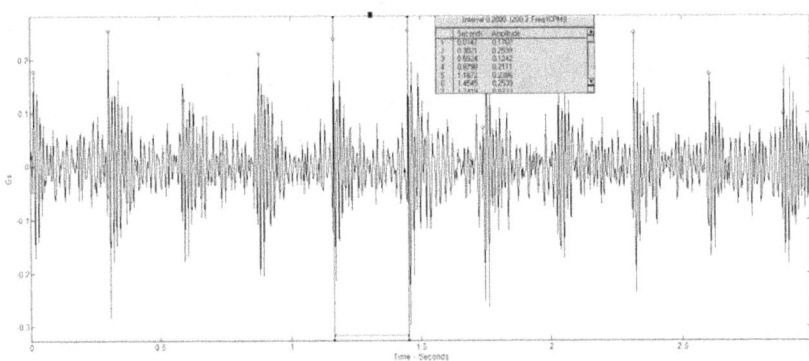

The clicks generated by the GM are shown as repeating spikes as the damaged area comes into mesh. Measuring between spikes gives an RPM of 208. Not exactly 225 RPM but close enough to verify which shaft has the damaged gear because the two RPM's are calculated using different mathematics. In the time domain it is a simple, F=1/t. In the FFT the frequency is calculated with a complex algorithm.

Gearbox Trouble Shooting

There are some mathematical factors about gearboxes that cannot be changed. The firm fact is that the gear mesh (GM) frequency is equal to the speed of the shaft the gear is mounted on times the number of teeth on the gear. And likewise, the gear mesh frequency will be the same if you use either the input shaft and it's gear or the output shaft and it's gear. Your first clue that you have a data error is when you arrive at two different gear mesh frequencies for the same gear set or as we found in this case, your collected data doesn't agree with your pre-calculations.

The usual error is in the number of teeth used in the calculation. It is often difficult to find any records of the tooth count on older equipment and the owner usually resorts to going to the storeroom and doing a physical count on the spare gearbox. Or, if there is not a spare, opening the installed box and counting the teeth.

This type of situation is illustrated by a case where we were asked to comment on a high-speed gearbox that had started making a lot of noise compared to the similar one along side.

An acceleration FFT will provide a treasure trove of information about what is happening inside the gearbox while it is in operation. Based on the information at hand, this gearbox had an input speed of 1490 RPM with a drive shaft gear of 125 teeth. The output speed was measured to be 7760 RPM via a 24-tooth gear. A routine calculation provided a gear mesh frequency of 186,250.

Examination of the FFT did not support the provided data. Note that the gear mesh frequency is shown on the plot as 147,263 CPM. There is an error somewhere and the FFT data provides a prime clue. Note that the 1X cursor is set at 1485 CPM. Then the sideband cursor is placed on the apparent gear mesh frequency and the heading says it is 99.17 orders of the 1X. This tells us that the input shaft does not have 125 teeth but instead has 99 teeth. (99 X 1485 = 147,015 CPM) and going the other way with the output shaft (147,015/7760 = 19 teeth) instead of 24. Slight variations in the numbers are from slight rounding errors, resolution of the spectrum and mathematical data collector processing.

Further examination of the FFT shows that the Delta spacing of the sidebands is 7673 CPM, which tells us that the signal is being generated by the output gear. If the spacing were 1485 then we would suspect the input gear.

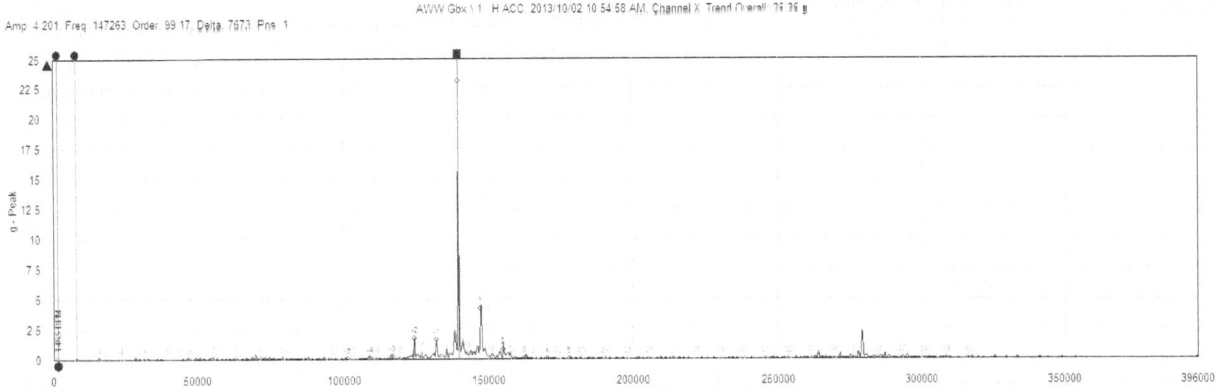

Across the top: Amp 4.201 Frequency 147,263 Order 99.17 Delta 7673

Although it was not done in this case, it is highly recommended that a time waveform spectrum be collected with each FFT on a gearbox. It doesn't take any longer and the data is there, it just needs to be saved. On a time waveform, the signal spacing and amplitudes will help confirm the data from the FFT.

Based on this data and the owner telling us that it was making a loud high pitch whine, we recommended the unit be shut down and inspected. We are still waiting for a follow up report.

Synchronous Gearbox

One of the best ways to be alerted to changes in your machinery condition is to trend your data. There are multiple programs that will do this and if used, will alert you either by a % change or an amplitude change in the flow rate, vibration level, electrical load, etc.

If you contract a consultant to evaluate your equipment, among the first things he/she will ask is to see the trend charts. Without this information all you are offering is a one-time look the situation, which limits the evaluation.

An example of this was the situation where the customer provided an FFT spectrum and a time domain spectrum of a critical unit with no historical data. Below are the two spectrums.

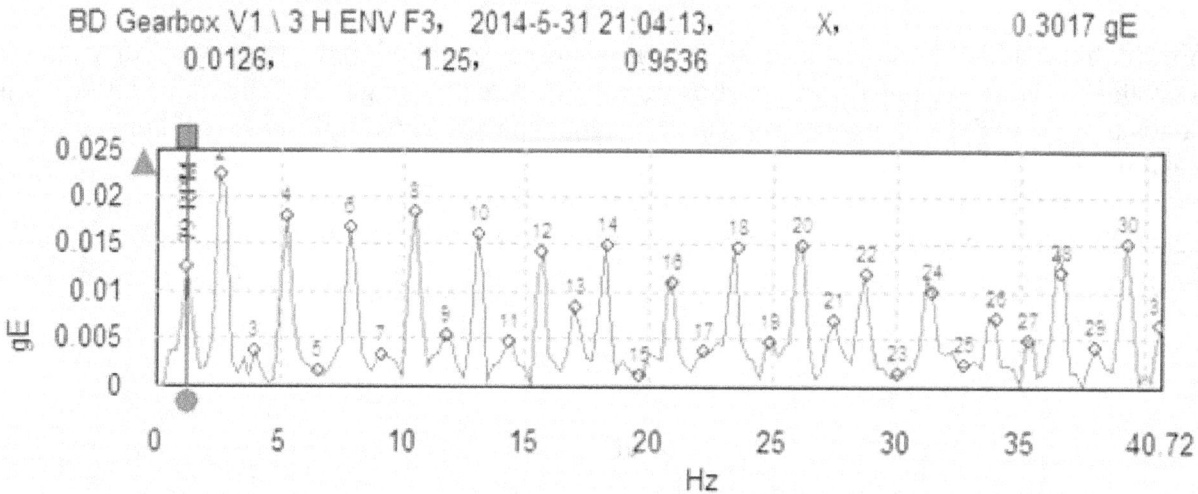

When using enveloped acceleration, gE, anytime you have multiple harmonics of the rotating frequency, 1X, you can almost always know there are problems. Although the amplitudes are low, so is the rotational speed and these amplitudes are significant.

When collecting data on a gearbox you should ALWAYS collect a time spectrum. From the FFT you can see there are problems but it is not clear what or where they are, it may just be looseness but you can't tell from just the FFT.

In the time spectrum notice the pattern of the signals, the rise and fall of the amplitudes. If the time between two peaks is measured, approximately 0.76 sec, this equals a rotation speed of 79 RPM so you then know that the vibration occurs once per revolution. Since this is a synchronous gearbox, both shafts are rotating at the same speed so a physical inspection will be necessary to determine which or both gears are in trouble.

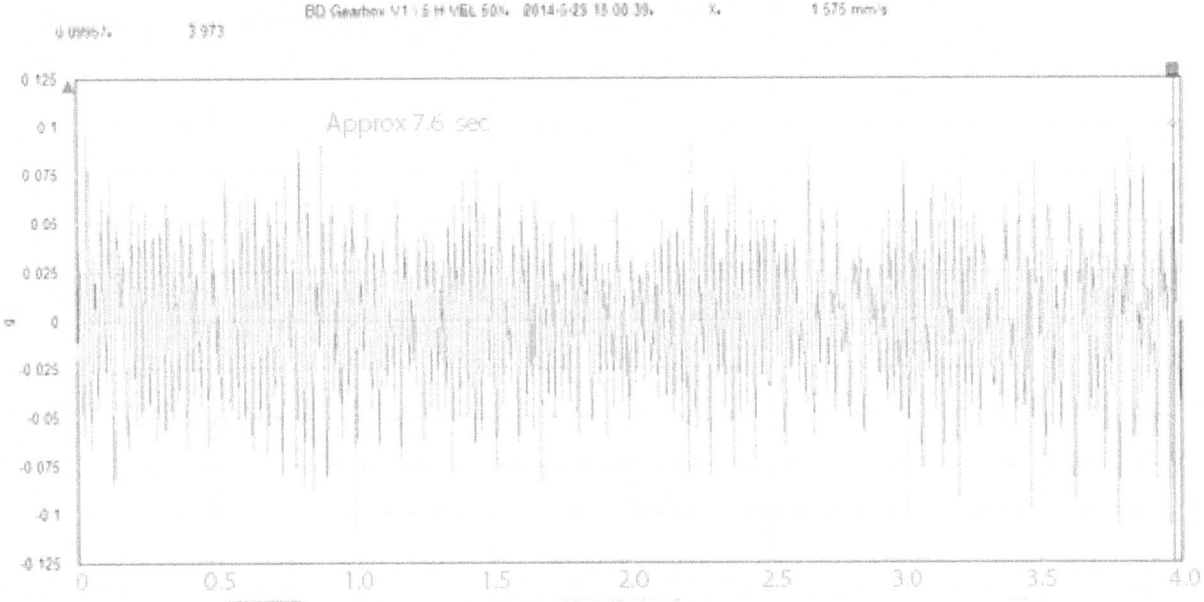

A physical inspection of the gears was ordered and this picture shows the damage that was found.

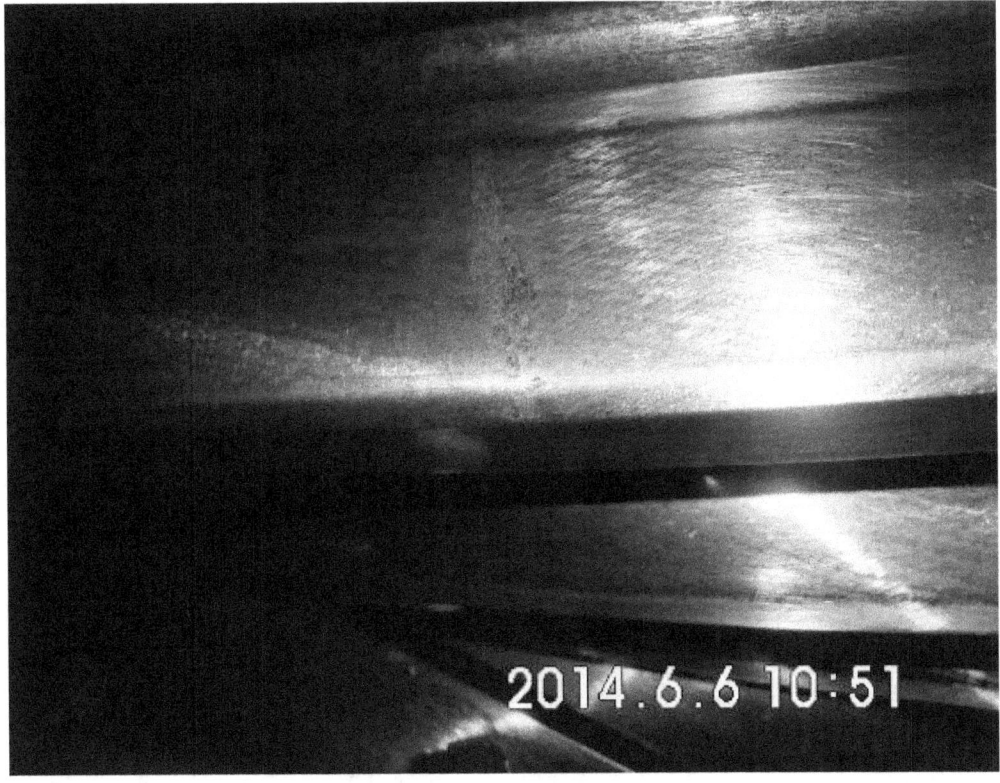

Early detection of the damage prevented an expensive failure and allowed the work to be done on a scheduled basis rather than unscheduled overtime.

General Maintenance

Motors

Adventures in Misalignment

Because misalignment is such a large factor in maintaining machine availability, it is a subject which need constant attention. Some machinery specialists estimate that 70-75% of bearing damage and failures occur because of misalignment. I've never seen a controlled study to arrive at those figures but my experience tends to confirm them.

Without getting into a lot of detail, here's what happens to the mis-aligned machine. For example with a motor and pump, there can be two types of misalignment. Parallel or angular misalignment. With parallel misalignment, the center lines of both shafts are parallel, but they are offset. With angular misalignment, the shafts are at an angle to each other. Either condition results in a push/pull force on the bearings, taking the load zone out of the bottom of the ring and putting it onto the side of the bearing, alternating sides with each rotation. In addition to damage from increased heating, the improperly loaded bearing will fail from the mechanical forces.

The cost of misalignment was well documented at a refinery where there were two identical vertical pumps that were periodically swapped out during operations. This facility kept very good maintenance cost records and after 10 years Pump A had incurred $200,000 more in costs vs. Pump B. It was well known that Pump A had a problem, it was nicknamed "Old Shaky", but no one had been able to determine the cause of the vibration.

We determined that the motor was vibrating because of a resonance condition. By placing a 4X4 against the motor housing and braced against the wall, we were able to apply a force that increased the stiffness of the assembly. As we applied the force and monitored the vibration levels, the amplitude decreased, when the force was removed the amplitudes returned to their former levels.

The motor was pulled and it was found that the original foundation was not level and plumb. The motor/pump was out of alignment and the condition had caused the continuous problem for 10 years.

Electric Motors and Bearings

A fairly common problem that occurs with electric motors is known as "soft foot". Under ideal conditions the motor is mounted on a base that is solid and flat. In practice, motors are often mounted on bases that are "almost flat" and sometimes even cracked. This results in several adverse conditions, bearing damage, motor housing damage and excessive vibration until something breaks. This would be a typical setup.

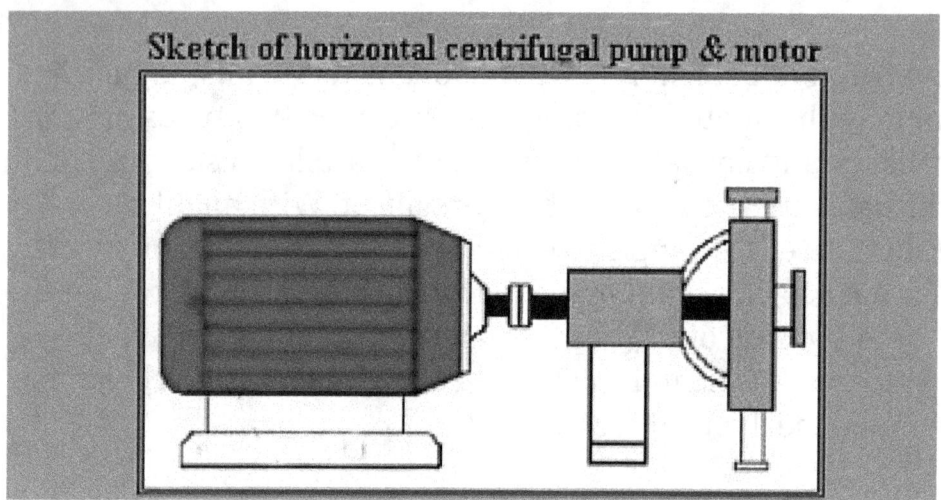

In an analysis it is common to check the amplitude of the frequency at 2X line frequency, 120 Hz with 60 Hz power or 7200 cpm, 100 Hz with 50 Hz power or 6000 cpm. A soft foot is one case where a mechanical fault can cause an electrical problem. It is common to see amplitudes of 0.04 to 0.10 IPS at 2X line frequency but higher amplitudes can mean that the rotor is not in the center of the magnetic field. In the area where the rotor is closer to the stator, the magnetic pull is higher. The stator's magnetic field rotates at the line frequency of 3600 cpm (U.S.) times the 2 events, the passing of the "N" and "S" poles, resulting in a 7200 cpm vibration.

A quick check is to monitor the vibration while the motor is rotating and loosen and retighten each foot in turn (if this is allowed by plant rules). If one foot causes a reduction in the amplitude and it returns when tightened, then that is the soft foot and it should be shimmed to place it in the same plane as the other three.

The warping of the motor housing from the soft foot can distort the bearing housing placing the bearing in a position to generate excessive wear, resulting in increased noise and reducing the life of the bearing. Because many plants do not do frequency analysis on all their motors, It has been our experience that this is usually the first sign of a problem and we pull up the old saying, "The noisy bearing gets the attention".

Electric Motors Part 1

The most common piece of equipment in most any plant is the electric motor and for most people all they know about electric motors is that you plug them in or wire them up and they rotate. If they stop rotating the usual fix is to call an electrician, which results in the installation of another motor and the old one is "sent out" to be repaired.

An electrical motor is made up of two main parts, the stator and the rotor. The stator is attached to the frame of the motor and consists of multiple insulated wires in the shape of coils. The rotor, the part that turns, is supported at both ends by rolling element bearings for motors from fractional sizes up to 200-300 HP range. These are the smaller motors; the larger motors are supported by plain or "sleeve" bearings. The stator is composed of multiple iron bars, the number depending on the size of the motor but in the range of 45 to 60 bars.

Industrial motors are generally powered by 3 phase AC or DC. A 3-phase motor is connected to 3 leads from the motor control center with each lead providing 1/3 of the power. This power, if observed on an oscilloscope, will be in the form of sine waves and is connected to the stator of the motor. When power is applied to the stator, a magnetic field is generated in the coils. Depending on how you wish to look at it, the rising and falling sine wave generates a force field that either pushes or pulls the stator as each lead is energized in turn. The speed of the motor is determined by how many coils are in the stator. With 60-Hertz (Hz) power a motor with 2 coils will rotate at 3600 RPM. 4 coils will deliver 1800 RPM. With 50 Hz power the RPM's will be 3,000, and 1500. These speeds are determined by the power and number of coils and on AC power cannot be changed.

There are two types of AC motors. Continuous duty, which once started run for days or "stop/start" motors such as on compressed air units which are cycled on and off as needed. These two motors are NOT interchangeable. The stop/start motor is much more rugged and if a continuous duty motor is substituted, it will soon fail from broken rotor bars as they are torqued on each start up. The mounting frame bolt hole pattern for two motors of the same HP rating is different for the two types to prevent the interchange, but determined people have been known force the fit.

DC motors are used in applications where it is necessary to vary the speed of rotation as in a paper mill or conveyor system. The basic DC motor is constructed the same as an AC motor but the power is different. If you examine a DC power source on the oscilloscope it will be a flat line not a sine wave. The plant AC power is processed through a Silicon Controlled Rectifier (SCR), which through the magic of electronics, chops the AC sine wave into many many tiny pieces and then puts it back together so that it no longer is a sine wave, but a flat line. But a flat power source will not make the rotor turn so the controller turns it back into a sine wave where the frequency is variable. So if you want your two-pole motor to rotate at 1800 RPM instead of 3600 RPM, you dial the line frequency down to 30 Hz and you get 1800 RPM. If that is too fast just dial the frequency back to 27 Hz and it will slow down to 1620 RPM. Smaller DC motors use the DC power and deliver it through a commutator. The commutator rides on the shaft, which has three electrical contacts. As the shaft rotates, the power is supplied to one contact at a time. When the next contact is energized, the polarity is reversed, producing the "push" or "pull" on the rotor resulting in rotation. This is the basic construction, there are variations depending on the application.

That's some very basic motor information. In Part 2 we will discuss some common problems you will encounter with electric motors, both AC and DC.

Electric Motors Part 2

In part 1 of this series, we discussed the general construction and types of electric motors, AC and DC. In this part we will follow the order of Part 1 and discuss the common problems that occur with electric motors and their controls.

All parts of the motor that involve insulated electric wiring, primarily the stator, are subject to damaged insulation, which occurs primarily from excessive heat. Motors have integral fans, which are designed to circulate cooling air through the motor, end to end. This air path can be blocked with debris, restricting the cooling air. Often production scrap will build up and this excessive "insulation" results in overheating and damaged wiring. Once the insulation is damaged an electrical short can/will develop creating additional damage that will spread exponentially resulting in a motor failure. On rare occasions portions of the wiring may be subjected to flexing which causes the insulation damage.

The rotor is constructed of a number of flat, iron or aluminum bars installed parallel to the shaft, which make up the rotor circuit. The number of bars depends on the size of the motor. Overloaded or defective bars will break at some point due to the torque generated by the current flowing in the stator. A broken rotor bar will reduce the speed of the motor because of the reduced reactive force. When one bar is broken, the bars on both sides have an increased load leading to additional broken bars. Usually with 3 or 4 broken bars, the motor will no longer start. Poor connections will increase the electrical resistance in the circuit and reduce the applied current, which will affect the motor performance. These increased resistance sources, broken rotor bars and poor connections, can be detected by performing a motor current analysis.

Industrial motors are generally powered by 3-phase power. The phase relationship between the three phases and the electrical current's amplitude must be balanced. Unbalanced electrical power will result in an electrically induced mechanical vibration. In a classic example, over the years an industrial plant was surrounded by housing developments. Eventually they began to notice that the vibration levels on their motors would go up in the morning and evening, the timing corresponding to the increased morning and evening activities in the housing area. Further investigations revealed that the power company had tapped one of the three phases to supply the private homes with power. When the residents arose in the morning and arrived home in the evening, the power usage of that phase would increase, pushing it out of phase and amplitude with the other two phases, resulting in an electrically induced vibration in all the motors. There is not an inexpensive solution to this type of problem and the plant had to tolerate it.

DC motors have the same problems plus one more. Because DC motors provide speed control via rheostat or frequency control, the possibility of operating the motor or the driven unit in a resonance zone is very possible. It is not at all unusual to see a warning sign on the speed control station not to operate the machine in a particular speed band. AC motors can be adapted for speed control using SCR's. However when the 3600-RPM AC motor is speed reduced and if it falls into a resonance zone, the user may wonder why his bearings have begun to fail at an increased rate. This is what happened at a large candy factory when their conveyor system was converted to variable speed with the addition of an SCR drive on the AC motors.

Electric Motors Part 3

This blog on electric motors concerns diagnostics when you become aware that the motor is not performing as expected. The primary instrument used in motor diagnostics is an electronic vibration data collector that is equipped to perform motor current analysis.

In general, when a motor is vibrating the usual suspicion is that something is out of balance without realizing that an electric fault in the motor can generate a mechanical vibration. By using the FFT portion of the data collector, the frequency of the vibration can be determined which will point to the source of the vibration. Frequencies equal to the rotational speed of the motor (1X) are usually associated with balance problems and a frequency that is twice rotational speed (2X) is usually associated with misalignment.

Motor current analysis will reveal if there is a higher than normal resistance in the rotor circuit. Note that this is the entire rotor circuit, including connections back to the motor control center although the usual source of the high resistance is one or more broken rotor bars. Without getting into the mathematics, (most data collectors have a built in motor current analysis app) the data collector will display the probable number of broken rotor bars. Physical symptoms of broken rotor bars are: slow startup, less than normal RPM, over heating of the motor and when the damage progress past 3 or 4 broken bars, the motor will not start.

Damage in the stator circuit is seen in an FFT with a strong signal at twice line frequency, (2LF), 7200 or 6000 CPM. Care must be taken with 2 pole motors to insure that the frequency is exactly 2LF and not twice rotational speed (2X).

One of the more common electrical problems with motors is a break down in the internal insulation because of excessive heat. Once the insulation breaks, the problem will only get worse over time until the motor fails. An infrared camera will quickly reveal a hot spot on the surface of the motor housing or if you run your hand over the housing the excessive heat can often be felt. When damage is present, a hot spot 6 inches in diameter and 20-30 degrees F hotter than the surrounding area can often be found.

Poorly constructed motors will often have the rotor not located in the center of the stator magnetic field. This causes points of stronger magnetic force at random points resulting in a push/pull effect on the rotor. There is no solution to this problem but to replace the motor and not buy equipment based on price alone. A good motor will exhibit the same vibration patterns if it is not mounted on a level base. An uneven base will lead to a "soft foot" because as the motor mounts are tightened the housing of the motor will be distorted which will distort the magnetic field inducing a mechanical vibration. If local procedures allow it while rotating, each foot, one at a time can be loosened. If as it is loosened, the vibration is reduced, then that is the foot causing the distortion and additional shims should be placed there to level the motor.

Finally, the biggest cause of motor vibration, in 70-75% of the cases, is misalignment. The analysis requires a data collector to verify with an FFT that the vibration signal is twice rotational speed and is primarily in the axial direction. It is predominantly in the axial direction because of the "push/pull" effect across the coupling, the results of the misalignment.

General Maintenance

Pumps

Detection of a Rubbing Impeller

As seen in this example, the speed of rotation does not change the resulting FFT when there is a rubbing component. The defining signal will appear at 0.5X where X= the rotation speed. Using enveloped acceleration will more likely show harmonics, as the enveloping process is more sensitive than velocity. As with any vibration, the higher the amplitude, the more urgent the need for repair.

The example is an FFT displaying the results of a rubbing impeller, and a time spectrum is also included. The time spectrum does not add anything to the diagnosis but verifies that the signal at 750 CPM is prominent. Based on the observed amplitudes, it was decided not to take fan out of service at that time but to increase the observations and trend the specific 750 CPM signal. The is acceptable as many minor machine faults will operate for extended periods without any problem, as long as they are monitored on a cycle in accordance with their criticality. The more critical the machine, the more often it should be monitored.

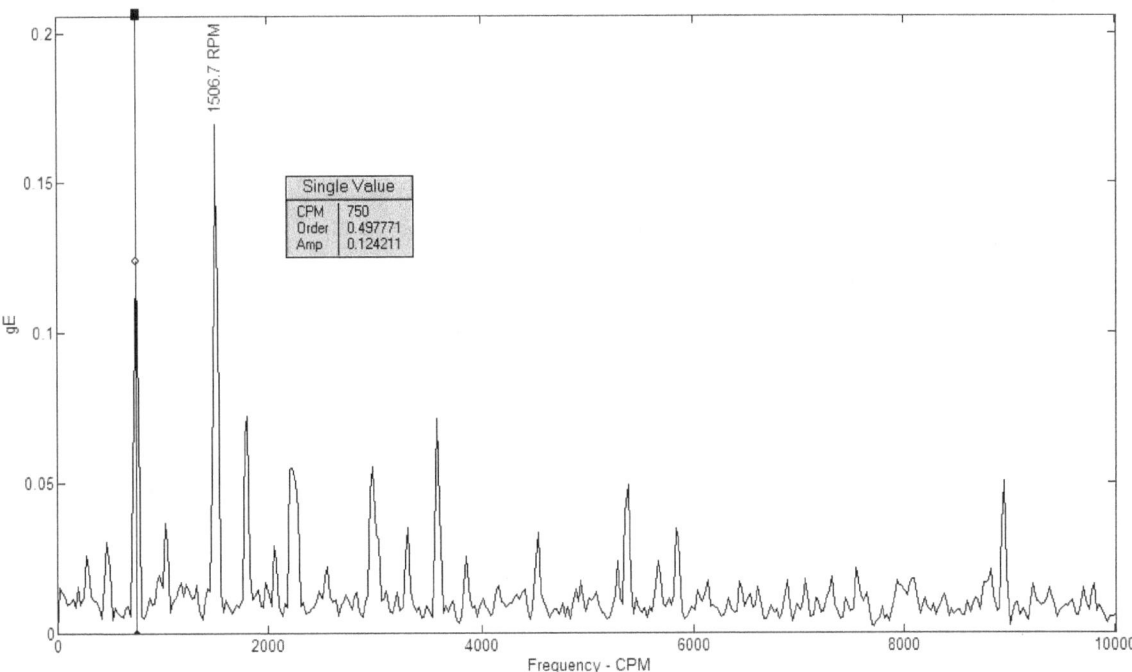

The data box indicates the element is rotating at 750 CPM and the frequency is 0.5X. This unit was operating on 50 Hz power and rotating at 1500 RPM. The operating speed of the machine is not significant; if a rub is present it will be indicated by a signal at 0.5X

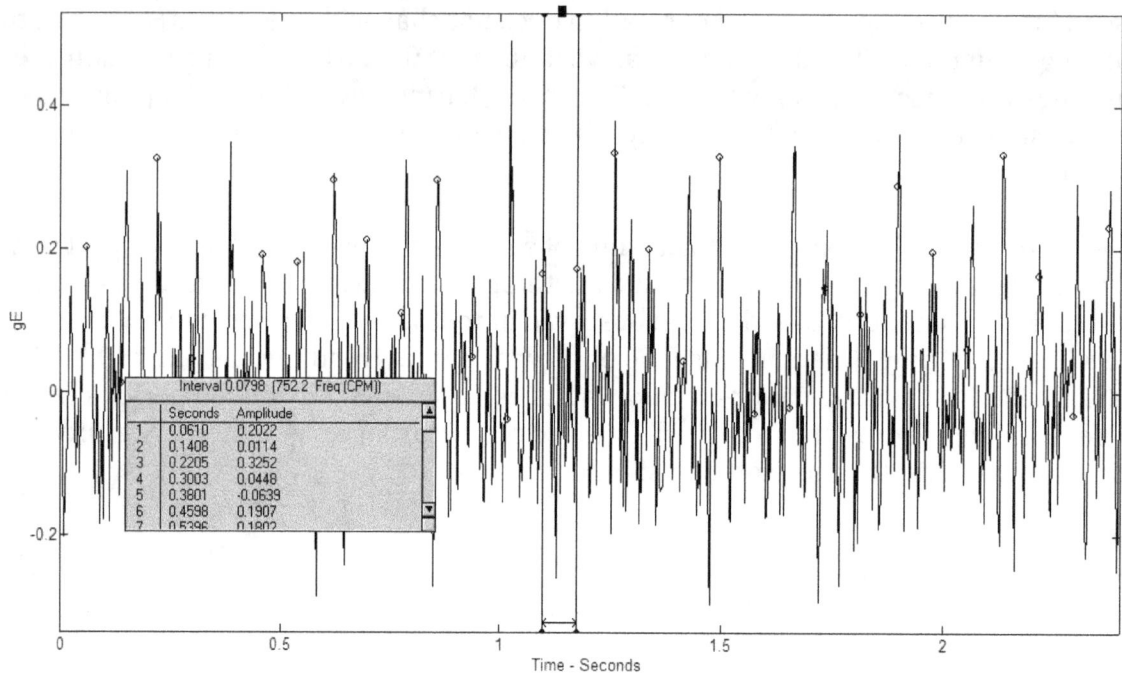

By placing the two harmonic markers on adjacent marked signals, the time lapse of 0.0798 seconds equates to a frequency of 752.2 CPM. (F=1/T) The accuracy of all measurements is dependent on the resolution of the spectrum and minor variations are not significant.

Misalignment

Although the general public thinks most machinery vibration problems are caused by something being out of balance, the general consensus of the vibration com-munity is that the root cause of 70 to 75% of vibration problems is misalignment. And, it is the root cause of 40-50% of bearing failures.

Machine failure because of misalignment usually proceeds in the following manner:

Every rotating element has some amount of imbalance. All manufacturers have a balance tolerance that they allow for during assembly. To obtain a near perfect balance would be an extremely expensive process, which is used only on such items as the super quiet equipment for nuclear submarines. This acceptable, small amount of inherent imbalance is dampened (absorbed) by the rolling element bearings. Bearing clearances between the rolling elements and races are approximately 0.0005", which is filled with lubricant. For all practical purposes, there is no clearance. As the unit is operated in a misaligned condition, the push-pull of the rotating elements, through the coupling, causes excessive un-designed loads on the bearings. Other than special cases, thrust bearings for example, the rolling elements of the bearing are designed to roll in the bottom of the ring, not up on the sides. These side loads lead to excessive clearances between the rolling elements and the races and the bearing is no longer providing the dampening needed to restrain the inherent imbalance in the rotating element. The final step occurs when some well-meaning person notices the vibration by placing his/her hand on the motor and writes the work order to repair or replace the unit, stating that it is out of balance. So, without a correct diagnosis, unnecessary work is performed. It would be much better to take the unit out of service for a short time and perform an alignment procedure and return it to service in an acceptable condition. As a sidebar, there are statistics that show that approximately 12% of any repair work done will have to be repeated for some reason, adding more to maintenance costs.

Also remember that just because a pump/motor, for example, comes to the plant already mounted on a skid, with acceptance stickers in place, doesn't mean that the unit is ready to operate. The alignment of the driver and driven unit must be verified. I have seen more than one such setup that was either dropped or bumped in transit and the two components were out of alignment.

Pipe Bound

One problem that sometimes occurs with both large and small piping systems is for the pump/motor unit to become "pipe bound". This condition is often detected during routine vibration measurements and it is noted that the vertical amplitudes are greater than the horizontal amplitudes. See the enclosed data block:

[mm/sec] (ISO 10816-3)

PUMP (vertical pipeline)											
1			2			3			4		
H	V	A	H	V	A	H	V	A	H	V	A
1.51	0.78	0.92	2.53	2.37	1.76	1.08	25.67	-	1.15	20.35	13.33

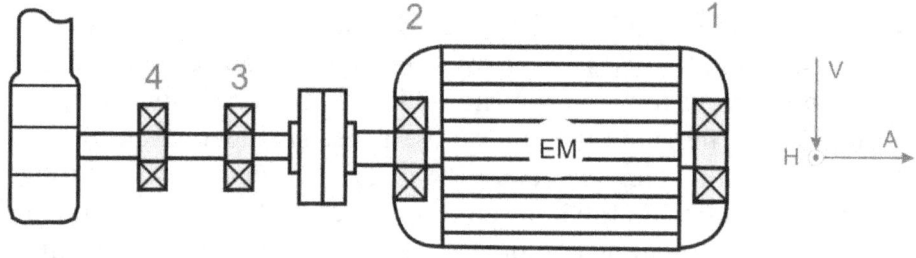

The high vertical measure is your first clue. Under normal circumstances the horizontal measurements will be higher than the vertical measurement because gravity and hold down bolts will prevent vertical movement.

The first action is to verify that the hold down bolts are properly torqued. If these are correct then shut down the system and disconnect the intake and discharge pipes. If the unbolted pipes move 50 to 75 cm then the system is pipe bound and the piping should be re-laid.

Pumps and Piping

One fairly common problem incurred during the installation of new pumps and piping is the condition where the pump is "pipe bound." Even with precise computer drawings, small differences in manual measurements can add up to the point where in order to get the new piping to connect to the pump, a come-a-long is used to pull the pipe into position. Then the connection is made and all appears to be fine.

The first time vibration measurements are made on the new pump; the resulting data is distorted because the excessive pipe forces don't allow the pump to vibrate in its normal manner. For example these two FFT's are horizontal and vertical measurements on the outboard bearing of the pump. It is not normal to have nearly "0" movement in the horizontal direction and over 25mm/sec in the vertical direction.

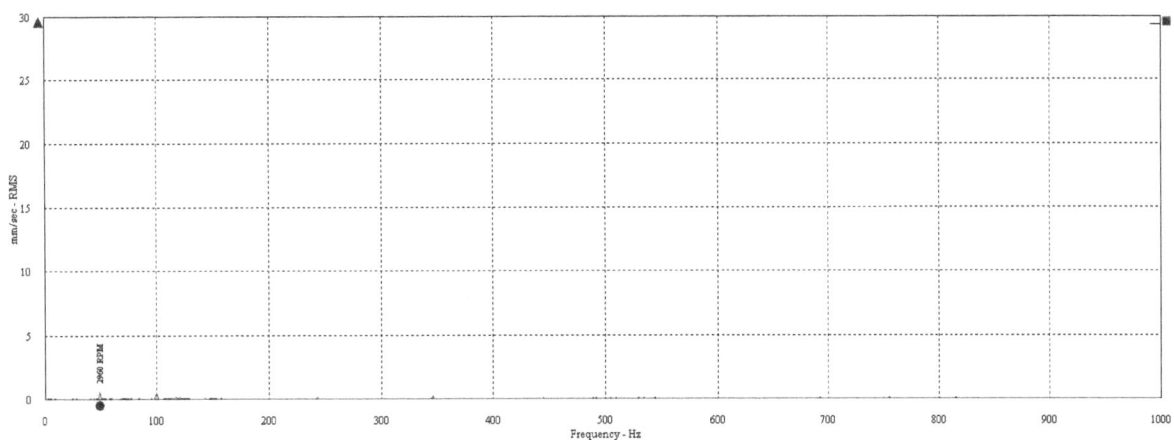

Horizontal

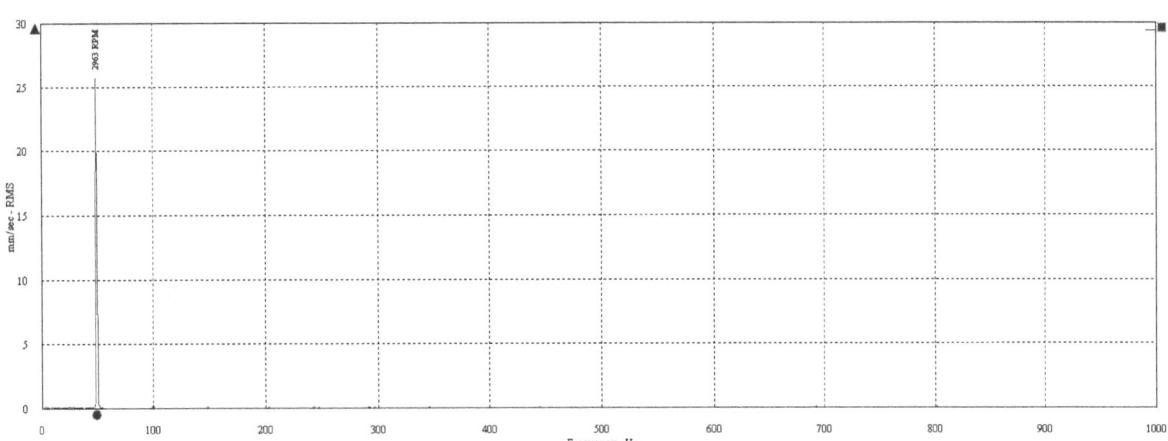

Vertical

The first response is to uncouple the piping. Generally ff the connecting pipe, of any size, moves more than 1 inch or 25 cm, the piping may be restricting the movement of the pump and should be realigned. Note that pipe binding can occur in any direction.

Another possibility in this case is that the hold-down bolts are loose or the foundation is faulty. Under normal situations the horizontal vibration is usually higher than the vertical. Gravity and hold down bolts prevent vertical movement but usually equipment can be excited in the horizontal direction unless there are additional supports. A good example is to grab the edge of your table or desk; it is difficult to move up and down but can be shaken back and forth fairly easily.

The FFT's shown were from a new installation where in fact, the piping had bound the pump. Several pipe segments were replaced and normal operations were restored.

Miscellaneous

Classified Maintenance Service

When dealing with government organizations there will be occasions where part of the maintenance service may involve classified equipment.

SKF Reliability Systems was contacted by a major military aircraft manufacturer and asked if we could supply the personnel and equipment to evaluate a signal they had recorded and wanted analyzed. From their description it appeared we could handle the problem, they had recorded a vibration signal from an installed compatible accelerometer.

On arrival at the plant the fun began. First was clearance through the main gate. Next was an escorted clearance into the building. Then it was an escorted entry through an internal building, down a long hall and a coded password for the escort with entry to a rather common looking lab.

The lab engineer explained they wanted us to analyze the recorded signal from a tape recorder setup on the table. We asked what kind of equipment did the signal come from, he responded, "Can't tell you that, it's classified". O.K., so we plugged the tape recorder into the Microlog and started the playback. We asked how fast the equipment was rotating? He again responded, "Can't tell you that, it's classified". O.K., since we knew what kind of airplane they manufactured at that plant, we asked if it was from a B-2. And you already know what he said, "Can't tell you that, it's classified".

There was a nice clear FFT on the screen; it had a number of what appeared to be harmonics of something rotating at 750 CPM. So we told him that whatever it was that was operating, that it appeared to be rotating at 750 RPM and based on the appearance of numerous harmonics of that frequency, there was looseness in the item or looseness in it's mounting.

With that, the engineer slapped his hand down on the table and said, "I told them that (blank) was loose, thank you".

With that he had us delete all the data from the Microlog and showed us to the door.

Total time flying to the site, getting clearances, getting back out and returning to the point of origination, 3 days. Total time performing the analysis: 15 minutes. Such is the life of a field engineer.

Consequences of Buying Cheap Products

On initial startup of a new machine rubs may occur because of misaligned bearing housings, an unbalanced rotor, misalignment with the driver, a distorted base, and the list goes on. In any event, the prime FFT indicator of a rub between a rotating element and some stationary portion of the machine is a signal at 0.5X. There may or may not be harmonics associated with this signal and they may also be seen at 1/5, 1/4, or 1/3 of 1X signal. Because so many submultiples of 1X are possible, the possibility of exciting a resonance is increased. A usual indication of a resonance is if any of the energy spikes in the FFT have a broad base and are surrounded by a cluster of small spikes as covered in the section on resonance.

The example is an outboard motor bearing on a new inexpensive motor. As often stated, you get what you pay for and this motor was unacceptable. When the motor was started, the vibration was excessive and it was shut down. When the FFT was taken, the rub signal was quickly apparent and the motor was removed to the shop.

On teardown it was noted that the outboard and inboard bearing housings were offset from the centerline so that the rotor was cocked to one side and there was a bright shiny area on the housing where the rotor hub had rubbed. The motor was discarded and the purchasing department told not to buy any more of this brand.

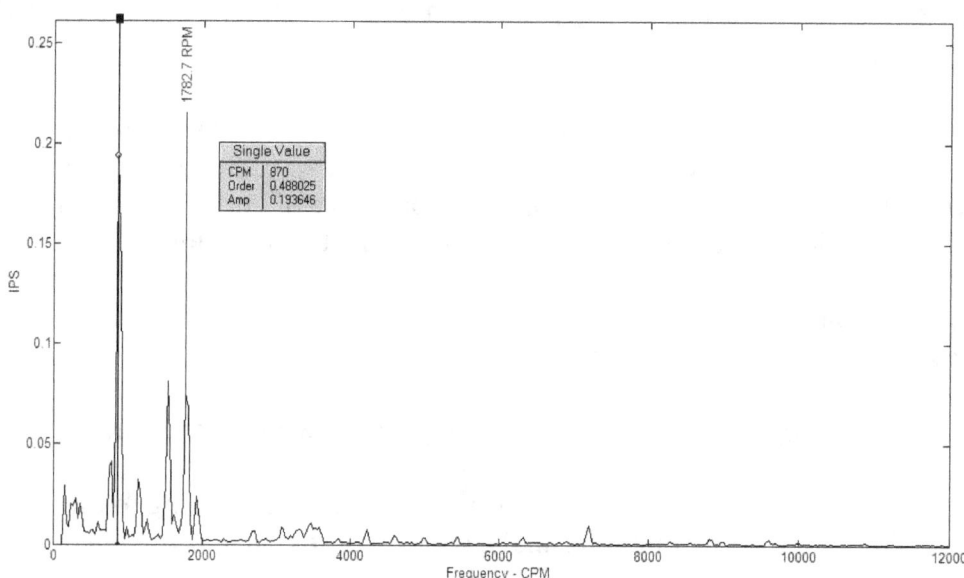

Duct Tape and Locktite

Maintenance personnel are often placed in the position of "Get it fixed, NOW". This pressure often seems to occur around 02:00 in the morning after a long day.

The initial vibration analysis was that the bearing appeared to be loose on the shaft; the FFT spectrum showed multiple harmonics of the rotation speed and the 4th harmonic was higher than the 3rd and 5th. Multiple harmonics of rotation speed is an indication of looseness and when the 4th harmonic is elevated, the bearing is loose on the shaft.

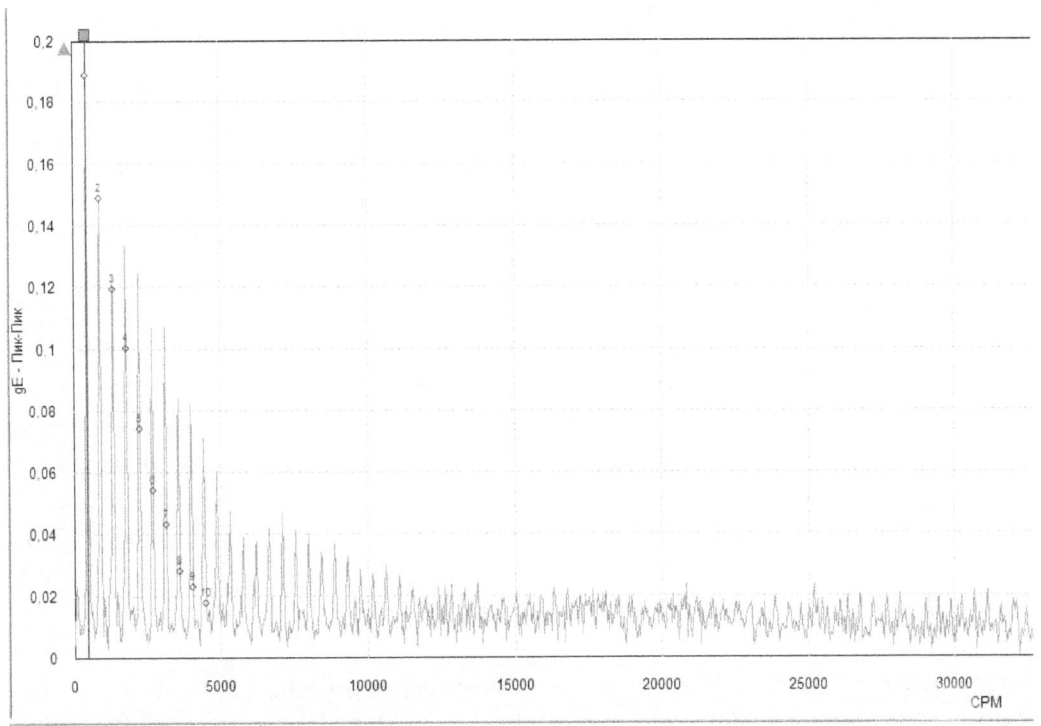

When the bearing was removed, an SKF 22248 CC/C3, the clearance between the shaft and the inner race was found to be 0.32mm. In an effort to speed up the repair, a new bearing was installed and Locktite applied between the shaft and inner race.

Unfortunately Locktite is not the proper fix for a worn shaft. One solution that will work is to remove the shaft, place it in a lath and apply a new surface to the shaft with a plasma torch. Then while the shaft is in the lath, turn it down to the proper size and reinstall.

And the duct tape? It was successfully applied to a crack in the fan housing that was leaking air. Several years later, the tape is still there and there is no air leakage.

Equipment Storage

The famous comedian, Bob Hope, once said that there are no new jokes, just new versions of jokes that are hundreds of years old. He was an expert so he should know.

In the same manner, except when a new technology comes along, there are no new maintenance problems. How many articles have you read concerning machine storage? I'm sure if you look back through maintenance bits, that you will find articles on machine, or machine components and their storage. Even with all that information available the following letter was recently received from an SKF European customer who related one of his current problems.

"... All I can add at this time is we too have had failure of equipment which has been in storage for a few years. An example is our conveyor pulleys have the bearing mounted and pulleys stored just sitting on the ground, with the bearing mounted in the pillow blocks and hanging on the shafts. These bearings are SKF 241/600 so a fairly large bearing and every one had false brenelling once installed 5 years ago. Since the pulley is not in a stand where one can come along and rotate it every few weeks the bearings are guaranteed to fail. Our last plant outage which is every 4 years we took all of our pulleys and had them shipped off site to have all bearings replaced, these were new pulleys with new bearings but stored improper, I believe that it is $100,000 per bearing plus two bearings per pulley and a total of 5 pulleys, and cost of labour and shipping, so this was a huge loss. Without proper procurement procedure you lose a lot of money. Another interesting fact is some vendors will ship equipment to you under "short term" storage procedures, short term is like 3 months. If you require longer storage then their storage procedure is much different. This detail needs to be included in the contracts and procurement part of your project. It is disappointing that we all know the equipment must be rotated and some require special storage stands but this never gets the attention it should."

This is a real life example. $100,000/bearing, 2 per pulley, and 5 pulleys. My math says just that cost was $1,000,000. I wonder who had to explain that to the plant manager!

So a word to the wise, make arrangements for the proper storage of your equipment and their components. Your job may depend on it!

Free Information

SKF offers several sources of information for the maintenance professional and to those who have general questions.

This blog provides rather specific information on particular subjects as you can see as you scroll through the various posts.

In addition, the reader can access more detail information at the web site; www.skf.com/cm/tsg At the top and side of the page is an index, which includes the SKF Forum. This site will provide you with over 500 case histories from all over the world. The Forum does require a free registration.

Check out these sources of information, someone once said that we have to learn from the mistakes of others, we can't live long enough to make them all ourselves.

If You Don't Believe Us, We're Outta Here

The problems that go with being a consultant are numerous. You've never been in the plant before, you don't know anyone and the big one is that they don't know you, just the reputation they've heard and accepted to justify paying the fees.

We were asked to assist a company that manufactured coaxial cable. There are a lot of types of coaxial cable; their product consisted of a heavy duty copper wire with nylon standoffs mounted every centimeter to maintain the proper clearance from the surrounding outer sheath. After the cable was completed it was wound on a reel to the desired length. For testing, a multi-frequency signal was applied to the cable while it was on the reel, and the output signal measured and compared to the input signal. Measured against industry standards, the cable was graded as either "A" grade or "B" grade based on the signal output.

For a number of years there were no apparent problems. They sold all their "A" grade to the more advanced users and were able to sell all the "B" grade product to less demanding customers. Then improving technology caught up with them and the demand for "A" cable began to exceed their output and the no longer desired "B" cable began to stack up in the warehouse.

As I understood the problem, when the copper wire was pulled through sizing dies and roller spindles, any vibration could "mark" the surface of the copper wire. When in service, these periodic marks would clip and degrade the signal. The usual suspected source of the marks were the spindles and the solution had been to change the spindles. That worked for a while but the "A" cable production began to fall behind and they asked for our help.

From a vibration standpoint it was a rather simple problem. From their data we knew the clipped frequencies and on the FFT we could verify that those signals were present in the spindles and were the source of the marks on the cable. This was our report to the production manager, which he rejected out of hand because, "Those are brand new spindles, they couldn't be the problem." Of course we had checked the complete line and were sure the spindles were the problem but, if you don't believe us, we're outta here!

About three weeks later we received a phone call and asked us to come back and do a complete survey of their other two production lines. Our initial reply was, "Why do you want us back, we told you what the problem was and you didn't believe us. We don't want to spend our time and take your money if you aren't going to follow up on our recommendations".

In a very contrite voice they explained that after we left they had gone over the complete line and couldn't find any source of the damaging marks. Out of desperation they had changed the spindles and mounts as we had suggested. The results were that the production of "A" cable had gone from 45% acceptable to 76%. They produced thousands of feet of cable per week and with a price differential of about $1.00 per foot, there were some very happy managers.

And that's why some consultants like to work for a percentage of the savings rather than a set fee!

Infrared Thermography

One technology that has many uses is infrared thermography. Two extremes is the scanning of structures to identify heat loss and insufficient insulation and the other is scanning suspected plots where a body may be buried. On a more practical usage by electricians is to determine the location of loose electrical connections from both inside and outside of the plant. In my experience at a nuclear power plant we used infrared to determine if the hydrogen evacuation nozzles in the ceiling of the containment building were cycling as programmed.

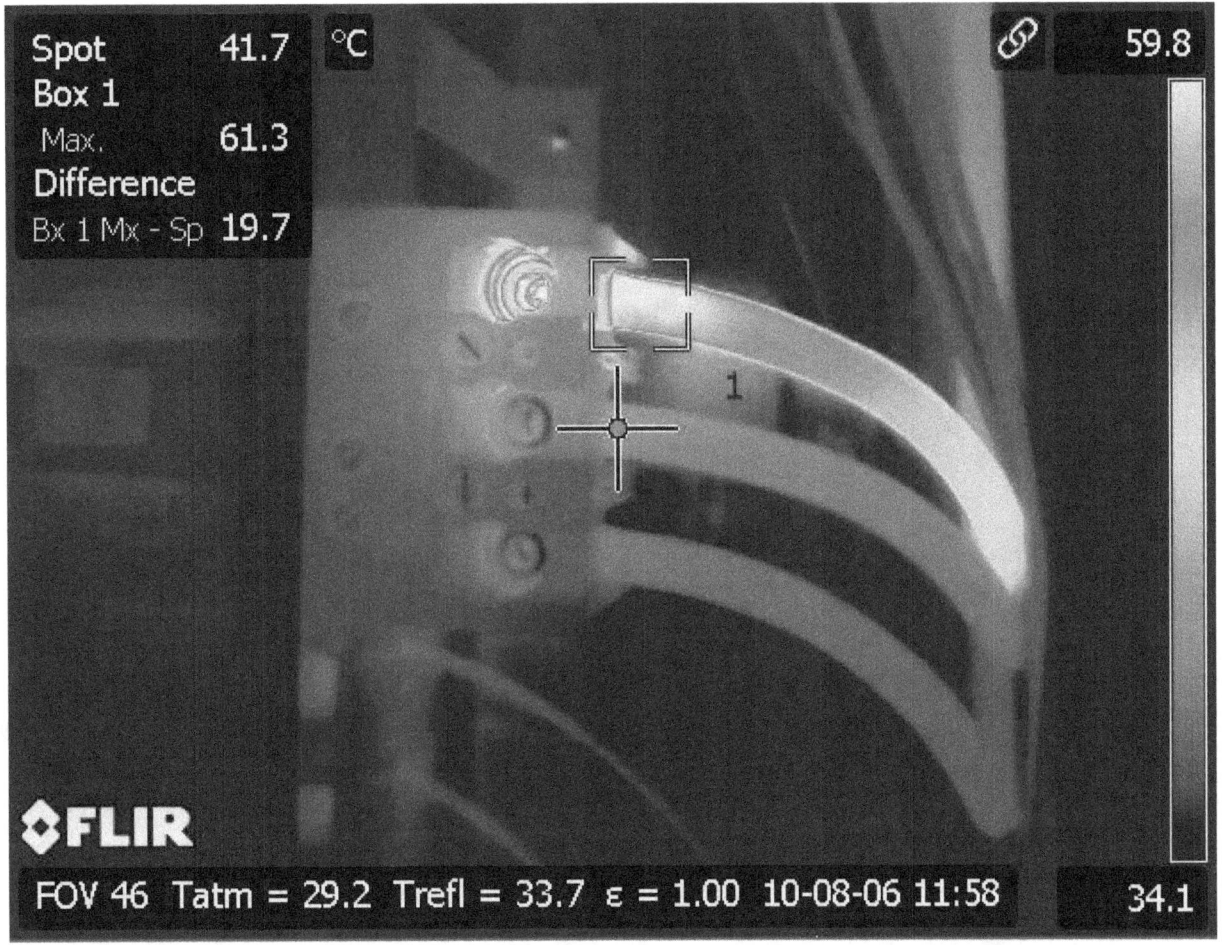

This example is a screen capture of an electrical connection where based on the color the top lead indicates a temperature of 61.3 degrees C and the other leads are at 41.7 degrees C. with a delta T of 19.7 degrees. (Ignore that the math is 0.1 degree in error)

If one were to look at these leads there would be no visible difference when seen by the naked eye. It takes infrared to spot the differences. Of course you might say you could feel the heat and know there was a problem but that determination would be much more difficult if the connection was on the top of a power pole!

To be done properly the operator should be certified and the necessary training and certification is usually provided by the manufacturer of the infrared equipment. Infrared is a valuable maintenance tool.

72

Keeping Good Records

One cause of bearing failure that often occurs and leaves few clues is the failure caused by greasing a sealed bearing. Over the years we have seen a number of cases where the vibration history of the bearing gives no clue as to what could cause the failure but when an autopsy of the bearing pieces is conducted, all we have is a jumbled mess of a twisted cage and loose balls.

The one common factor is that these bearings were sealed bearings. And the purpose of the title of this blog is, do you know which bearings on your machines have sealed bearings?

What often occurs in manufacturing plants is that the newest member of the maintenance group is often given the task of checking the grease and oil levels on the equipment. Minimum training, which usually consists of telling the person where the lubricants are stored and such broad instructions as "keep the oil level to this line" and "just put two squirts of grease in each bearing" is all that is provided.

If the lube "specialist" is like most folks, then if two squirts of grease is good, then three or four will be better. WRONG.

If they are sealed bearings, then they do not need ANY additional grease. They are filled and sealed at the factory and any additional grease is both wasted and damaging. Damaging from the standpoint that grease guns can develop over 1,000 PSI and this will force the seal inward and interfere with the rolling elements. Shortly afterwards the bearing fails and the inspection shows there was plenty of grease, and everyone wonders why the bearing failed. The machine records must show whether sealed or open bearings are installed and this information must be made available to the lube specialist.

The other reason that three or four squirts of lubricant are not good is that even for open bearings, that require periodic greasing, any open bearing is designed for the certain amount of lubricant. Any excess is wasted and will cause the bearing to overheat and be damaged. When the bearing is greased a measured amount should be applied on a scheduled basis.

This is a picture of a typical sealed bearing. The seals may be made of different materials depending on the service required but in any case you cannot see the rolling elements and they do not need additional lubricant.

Leak Detection

Vibration data collectors are becoming rather common in the maintenance field. But did you know that they can also be used for leak detection on valves and steam traps?

When a gas or liquid passes through an orifice, a vibration is induced into the pipes. This vibration covers a wide frequency spectrum. However, because it is generally in the higher frequencies, it will attenuate (fade) as it moves down the pipe. Testing has determined that a leak-generated signal will attenuate in approximately 10 pipe diameters. For example, on a 2" line, the vibration would attenuate within 20 inches of the valve.

The procedure is to mount, with a magnet, an accelerometer on the valve stem of the valve. This is the ideal site because the valve stem is connected directly to the valve plate. If it is not possible to reach the valve stem, mount the sensor on the valve body. Caution: The temperature of a steam line will often exceed the temperature limits for the accelerometer. In these cases, a steel (never aluminum) stand off rod should be used to prevent the destruction of the sensor. A reading is taken using the acceleration or acceleration enveloping mode and the overall amplitude noted. Next a reading is taken at least 10 pipe diameters down stream of the valve with the same setup i.e., if a standoff is used, it must be used for both readings. This second reading provides a background noise reading. Compare the two readings and if the reading from the valve is 20 to 25% higher than the background reading, the valve is probably leaking. For example a gE reading of 1.0 on the valve and a 0.70 background reading would indicate a leaking valve.

This procedure was used in a plant with 26 valves, which were required by regulation to be examined for leakage. The valves all drained into a large manifold and which fed to a storage tank. If a valve was leaking they could tell by the level controls on the tank but they had no way to determine which of the 26 valves were leaking. The prior procedure was to rebuild a portion of the valves at each outage. They would start at the beginning of the shutdown and work on suspected valves until it was time to resume operations.

Using leak detection, while the plant was in operation, a set of measurements was taken on each valve and line then ranked high to low on a chart. Maintenance was instructed to start at the top of the list and work their way down the list until the outage was over or valves were found to have no leak indicators and then stop. After several cycles, the amplitudes that were determined to be background noise were documented and no more valves were opened that were below or equal to that amplitude. This saved many man-hours of unnecessary work and allowed the crew to go to other jobs. As a side note, historical data shows that overall; maintenance will have to return to 11% of their overhauls and repairs to redo the work because of a pinched gasket, missing bolt or some other mistake. A wise man said, "If it ain't broke, don't fix it!"

Leaking Valves

One of the systems in nuclear power plants is the piping arrangement for the distribution of borated water. And as was explained to me, in case of an accident the borated water is released to mitigate the radioactivity. In order to be immediately available the system is under pressure when the reactor is operating. In case there is an emergency the system is discharged as designed. Because the system is constantly under pressure, there is a possibility that any of the 26 valves in the system could develop a leak.

Another aspect of the nuclear power business is the concept of known leaks and unknown leaks. This particular system piped all the valves to a common holding tank, which had level monitors. If the level increased there must be a leak. Now, which valve is leaking? As with many plant systems, nuclear and non-nuclear, if you can identify the volume discharging through a leaking valve, operations can continue to a higher limit but if the volume or source is unknown, the allowable leakage limits on the holding tank are lower and the plant must be shutdown.

Acoustic monitoring provides answers to both situations. In restricted areas, acoustic accelerometers were permanently mounted on the valve stems. After a baseline measurement is established the amplitude of the acoustic signal is trended. If the valve develops a leak the acoustic signal will increase. With all data collected at a constant temperature and pressure, the resulting trend of the acoustic signal amplitude will be a linear response. Then using a chart developed in the lab at the same temperatures and pressures, the amplitude is entered on the "X" axis and the volume is read on the "Y" axis below the intersection of amplitude and the lab produced linear response plot.

In the case of the 26 valves the previous preventative maintenance procedure was to rebuild 5-6 valves on each shutdown working in numerical order from previous shutdowns. This is really not very efficient because perfectly good valves were being rebuilt for no reason and valves in need of repair were missed. And in general, history has shown 11% of the time, when maintenance is performed; some component has to be redone.

The solution was to take acoustic measurements on each valve during operation. The valves were then ranked in descending order of amplitude and the top 5 or 6 were rebuilt. Within a few cycles the leakage problem had gone away and by rebuilding only the top ranked valves, satisfactorily operating valves were ignored.

It was estimated that by the time the valve rebuild program was fully in place it was saving one to two days of outage time on each shutdown. A conservative estimate of the cost reduction of one day saved was $2,000,000.

Lubrication

In any bearing the purpose of lubrication is to prevent contact between the rolling elements and the races or the shaft and a plain bearing. The lubricant is necessary, being composed of the proper additives and of the proper viscosity. The correct lubricant to be used is determined by the speed and load carried by the bearings. Other than special cases, grease and oil are the most common lubricants. Your SKF bearings dealer can provide you with the data needed to determine the proper lubricant for your application.

Viscosity is the prime determination in the selection of a lubricant. High or low temperature and speed determine the required viscosity. The lubricant forms a thin film between the races and the rolling elements. Insufficient viscosity will allow the two surfaces to make contact generating heat, wear and surface degradation.

Once a lubricant is selected it is just as important to use the proper amount. Excess lubricant will generate heat and can cause damage to the bearings. Bearing cages are not designed to plow through grease. In the case of electric motor bearings, excess lubricant will often escape into the motor housing causing a short in the electrical components. Some studies have shown that more bearings are damaged by excessive lubricant than those that lack a lubricant. SKF will also provide you with data on the amount of lubricant needed, based on your application.

Storage of lubricants is very important. If an oil storage drum is exposed to conditions where water can mix with the lubricant, degradation will occur. 1% water in the lubricant will reduce the life of a rolling element bearing by 90%. The bearing steel will attract the hydrogen molecules from the water, which will cause that spot in the steel to become brittle and flake off. These flakes then rust and contaminate the lubricant. As time passes, a spall will develop at the damaged location leading to bearing failure. If the lubricant is stored where it is exposed to dirt and other particles, these particles will be embedded into the surface of the ring and over time degrade the bearing to the point of failure.

Remember, your bearings are precision instruments. They should be handled and cared for with appropriate care.

Machines - Parts

Quite often it is the little things in life that rare up and bite us when least expected. We were working in a factory that makes large grinding machines. The price of the machine depends on the size and the one we were looking at sold for USD$250,000.

After changing the spindle, the pulleys and the belts, the quality control engineer was at his wits end trying to locate the source of a spindle vibration using a vibration pen. The amplitude on the spindle exceeded what was allowed before shipping. We were there for another project but he asked if we could take and look and see what we could find.

First we took a vibration measurement FFT on the spindle housing. This told us that the vibration was at 1780 CPM with an amplitude of 3.1 mm/s. Because the spindle rotated at 3800 RPM we knew that the vibration did not originate with the spindle but from another source that was vibrating at 1780. Of course the only thing that could be operating at that frequency was the drive motor. There was some confusion since we were in Europe, which operates on 50 Hz power, and what we were seeing didn't match with any 50 Hz sources. That was finally cleared up when they told us this machine was shipping to the U.S. and they were operating it on 60 Hz. Some days get frustrating!

Since the vibration was at 1X of the motor, we said the motor or pulley was out of balance. They had the standard answer, "Couldn't be, it's expensive and brand new". So we suggested they remove the belts and run the motor solo. Same results, the unit was out of balance. Next step was to remove the pulley and run the motor again.

We observed as the pulley was removed that it clamped onto the shaft with a set of screws around the perimeter of the pulley and had no keyway. However, the motor shaft had a keyway cut for installation of a key even though in this type of installation, no key was needed.

BINGO!

The amount of metal removed from the shaft for the key, which wasn't there, was enough to induce an out of balance condition. Solution, install the bottom half of a key in the shaft slot and reinstall the pulley.

Problem solved, the machine was well within vibration tolerances and was shipped that afternoon. Moral of the story, when trouble shooting, you've got to watch everything that goes on, even the little parts!

Detecting Mechanical Looseness

Even the most carefully assembled machines can develop looseness over time. Nuts work themselves loose, pillow block bolts break, shafts become worn and undersized, grouting breaks down and occasionally a wrong sized component is installed, allowing excessive clearances.

The predominant feature of an FFT of equipment that has loose components is a display of multiple harmonics of running speed. The rotating speed of the equipment is referred to as 1X, and in the following example is 1781 CPM. The second harmonic (2X) would then be at 3562 CPM, the third (3X) at 5343 CPM, etc. The spacing between each harmonic is equal to the shaft rotation speed, indicating mechanical looseness. It is not unusual to see harmonics extending to 10X and higher. Some texts will refer to 1X, 2X, etc. as "orders" where the nomenclature 5X would be referred to as 5 orders of running speed. Although the amplitudes of the harmonics vary because of the phase relationship between the harmonics, there is no particular interpretation that can be attached to the amplitude of an individual or group of harmonics in any analysis except in special cases of looseness of bearings. It is possible to determine to some extent the source of the looseness by taking amplitude measurements at several locations on the machine where the amplitude will be higher the closer you are to the source of looseness.

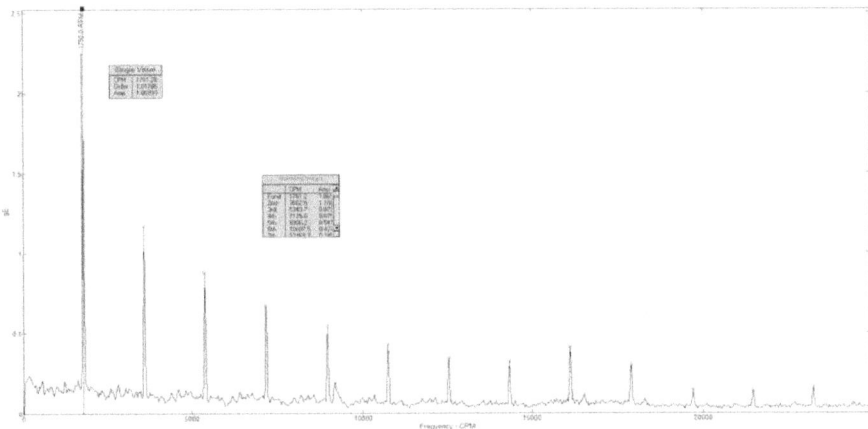

This FFT spectrum was taken on an exhaust fan where a base bolt had broken resulting over time in the looseness of the other three bolts.

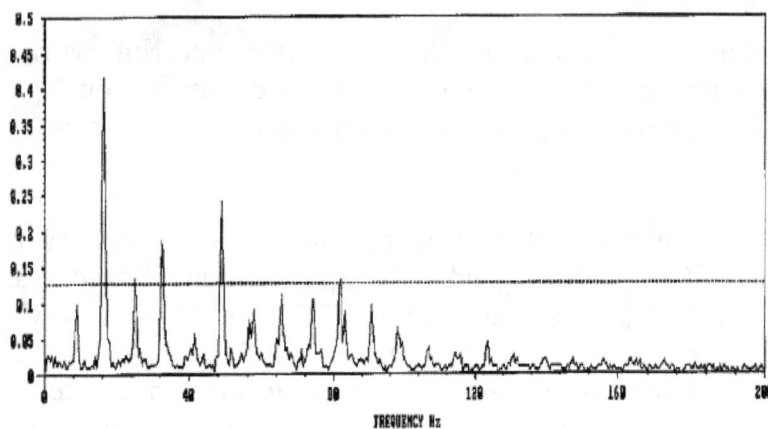

When dealing with plain bearings it is possible to generate multiple harmonics when the bearing has become worn and is not properly supporting the rotating shaft. When this occurs the shaft will be loose in the bearing and can move in response to outside forces such as balance and operational processes. In cases of extreme bearing wear, the shaft will roll up the inside of the bearing and since the oil wedge cannot provide the proper support, the shaft will drop to the bottom of the bearing and then roll back up, build the oil wedge and then collapse again.

Slow speed rotor with a worn plain bearing shaft.

Operating Vertical Equipment

A fairly common vertical installation is a pump submerged in a fluid and the drive motor mounted on the foundation above with a drive shaft that descends to the pump. These vertical installations can develop some unique problems.

Starting with the motor, it usually has bolts anchoring the frame to the floor. If you've ever balanced a pencil on the tip of your finger you know that it is much easier to balance the pencil in a horizontal position vs. a vertical position. The vertical drive motor is in the same situation. In addition to having bearings designed to carry the vertical load, it has a tendency to oscillate around the floor mounting. If the floor mounting is not plum and level, it is easy for a resonance to develop. One of the first things to be done with a vibrating vertical motor is to check the security of the mounting bolts then do a bump test on the unit. If there was resonance present, the bearings should be monitored, as resonance conditions are very detrimental to bearings.

Very long drive shafts should have supports to prevent shaft whip. The number and type of support will be determined by the shaft diameter and RPM. Often all that is needed is a sleeve to maintain the shaft position although in some cases this intermittent support contains a shaft bearing.

At the bottom the pump is subjected to the usual forces with the addition of being submerged. A common installation of this type is a fire pump submerged in the river to provide fire support. Because of the rise and fall of the river depth during the seasons, long shafts are needed. Installations using tidal water have the same requirements as the tides raise and lower the surface level. Ideally the pump will be mounted on some type of foundation to prevent movement. And finally the pumps may be operating in a hostile environment such as the river or ocean bay.

Remote Leak Detection

Inside many operating plants there are locations that for safety reasons cannot be occupied. However, there are fluid systems that contain valves that over time will begin to leak and the operator needs to know a leak is occurring. In a nuclear plant there are hundreds of valves, many inside the containment structure. One such system had valve drain lines that would route any leakage to a common tank. If the level of this tank began to increase, then the operators knew that a valve was leaking. But which one?

If the content of the system, in this case water, remains at a constant pressure and temperature, and a valve in the system begins to leak, it was proven in the laboratory that there is a linear response between the volume of leakage and the amplitude of the signal. Multiple tests were performed using operational conditions and the data was plotted to determine the slope of this response curve. At the next outage accelerometers were stud mounted on each valve body with the signal wires routed to a safe area.

In the nuclear plant the limit for an unknown valve leak was 1.5 GPM. However if the specific leaking valve could be identified then the allowable leakage was 2.5 GPM. After instrumentation each valve could be monitored during operation and if it began to leak, using the developed leak vs. amplitude chart, Figure 1, the volume of leakage could be calculated. This allowed the plant to continue operations until the leakage reached the limit of 2.5 GPM. Although this seems to be a small matter, it allowed the plant to operate much longer between shutdowns, which greatly increased its efficiency. When a down day costs in excess of one million dollars, efficiency is very important.

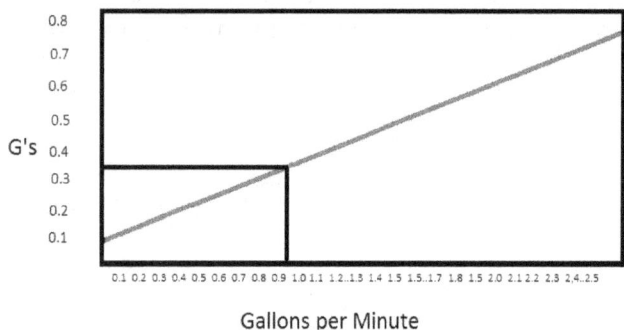

The same principal applies to any fluid, which includes air and gasses.

Return on Investment for Preventative Maintenance Activities

A discussion that has been going on for many years is how do you calculate the return on investment for preventative maintenance activities. There is no question that a maintenance staff is necessary, but what are they worth at the bottom line?

How do you prove your worth when you work on a machine that is scheduled for an overhaul but is still functioning properly. If it didn't fail, how can you prove you saved so many production units or you saved an unplanned plant shutdown? You have to have an acceptable metric to use as a baseline in order to arrive at a reasonable calculation.

One metric that has been around for a number of years, based on actual costs, and is generally accepted, is that the cost to repair a machine that failed is 10X the cost of shutting down the machine under controlled conditions and making the repair. The higher costs are attributed to: cost of corollary damage, increased overtime for call-ins, loss of material in process, safety of personnel and other such items that may be different for different processes. If those in charge are reluctant to use a factor of 10, use 8X or 7X, whatever you can agree on.

When I worked on the maintenance staff in a nuclear plant, we used this 10X metric in the following manner. The time cards are an official document. If we tagged a machine to be balanced, aligned, bearings replaced or the foundation repaired because it was outside of acceptable operating limits, then the time required to perform those functions was recorded on the time card of the persons doing the work. If for example it required 8 man-hours to perform an alignment on a motor/pump unit, then it was projected that if the unit had failed, it would have required 80 man-hours to perform the repair. This results in a savings of 72 man-hours. If the average wage rate is $50.00 per hour then the maintenance savings equaled $3600.00.

Using this criteria completely avoids any argument with the accountants over how much production was or was not lost, the avoidance of corollary damage or any of the other cost factors that might be involved.

At the nuclear plant, myself and two aides performed all the vibration analysis requirements. After getting a buy-in from management and using the 10X metric, we accounted for over $600,000 in labor savings in one year. What is generally a choke point is that most people do not like to blow their own horn, but I can guarantee you that if you don't, no one else will blow it for you. And the monthly report we submitted went right on up the line. And it affected us directly because when layoffs occurred, others were gone but they didn't touch us. We had proved our worth.

Risk

A fairly common question we get on the SKF Reliability Discussion Forum concerns whether to shut down the machine or let it run. And that is an open-ended question. The decision to run or repair involves safety, operations, parts availability, manpower and the list goes on with every case being different. All the mentioned factors involve risk. Evaluation of risk itself is also based on many factors and generally safety and cost are paramount.

We were asked to evaluate a furnace ladle bearing in a steel plant. To change it would mean lost production and a cost of $250,000. We were the second company they had consulted but the first company would not give them a decision, expressing the desire to collect more data. We evaluated our data and determined that the bearing had probable internal looseness from excessive wear and recommended the bearing be changed. When asked what could happen if the bearing wasn't changed we replied that they might dump a few tons of molten metal on the floor and anybody standing nearby. Did they want to take that risk? They changed the bearing and not only was the bearing seriously damaged, other internal parts of the tilting mechanism had locked up, resulting in the damaged bearing.

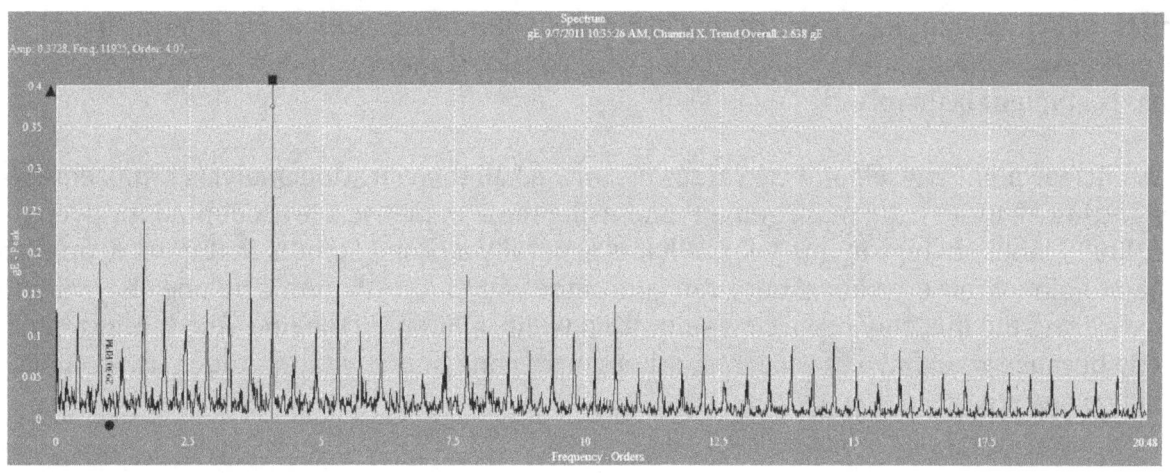

Typical response for looseness showing multiple harmonics of the rotating speed, using enveloped acceleration.

Case #2 was similar in that they couldn't decide whether to shut down or keep running a slow-speed conveyor belt system that moved sand from the crusher to the storage area. In Texas they crush rocks to make clean sand. Shutdown would certainly stop production but that was the only concern. If the bearing failed the only consequence was that the belt would come to a halt. If it overheated prior to failure a fire was not likely and frequent inspections could ease this worry. The cost to change the bearing would be the same before or after the failure and with the frequent inspections corollary damage was not a concern. The decision was to maintain operations. Three months later during a slack production time the bearing was changed with no lost production at a budgeted cost.

Remember to consider all factors in a risk decision.

Run to Failure

What is Run to Failure (RTF)? Of course the simple answer is that you operate the equipment until it fails, a philosophy based somewhat on the age-old statement, "If it ain't broke, don't fix it." The main problem with this attitude is that it often ignores the consequences of the failure. In fact, there are cases where run to failure is the best practice. However, it should be a deliberate decision and one should consider the following factors before proceeding.

1. Safety. If the machine fails is there any probability that someone might be injured? Safety should never be compromised.

2. Collateral Damage. The failure of a small horsepower motor may be of no consequence but if the failure of the motor interrupts the production process resulting in a raw material loss, then the costs go up. Also the failure of one component may cause the failure of another.

3. Cost of condition monitoring. Labor and equipment costs are in this factor. In some cases it just may not be worth it to monitor the equipment. Roof fans and such like come to mind.

4. Cost of repair. Where machines are tracked as cost centers, historical records will provide the data to again decide if it worth fixing or just let it run. Empirical data has shown that as a rule of thumb, it costs 10 times more to repair a failure vs. scheduled maintenance based on monitoring.

5. Cost of replacement. Does the equipment have a spare on standby or in the warehouse? Is there a loss of production during the switch over? How long does it take to obtain a spare or repair the damaged unit and how does that effect production?

These are the major decision factors that one must evaluate before running a machine to failure although certainly not all-inclusive. Others may have additional factors depending on their unique circumstances.

Shop Practices

Rolling element bearings are such common items in the maintenance environment that they often become just another item to handle. In fact, they are not a common item; they are a precision instrument that is needed to be maintained in a precise manner to utilize their full potential. It may be acceptable to leave wrenches and screwdrivers laying on the bench during the lunch break but an unwrapped bearing needs to go back in the box, unexposed to the atmosphere of the shop.

A large, well known, manufacturer called into the SKF service department and said that new bearings that were used in their infrequently produced "quiet" motors were making a buzzing noise during assembly line testing. The assembly process was observed and it appeared that they were using standard practices. Then it was noticed that multiple bearings were delivered to the line completely unwrapped and exposed. When questioned about this practice it was explained that this was their procedure, the bearings were in the storeroom, unwrapped and stacked flat on the shelves. The runners just went and collected a quantity when they were needed for the day's production.

The city where this occurred was well known for their poor air quality. An investigation and analysis of the buzzing frequency revealed that it matched the Ball Spin Frequency, indicating a fault in the rolling elements. A teardown of the bearing found that the rolling elements each had a small pit, which would match up with a pit on the side of the race. The conclusion was that over time, in a storeroom without temperature controls, an acidic condensate had formed and the moisture had corroded the two surfaces. When the bearings were mounted vertically, no contact was made with the damaged ring but the damaged rolling elements generated a buzzing sound as they rotated. If the bearings had been mounted horizontally, as they had been stacked on the shelf, we would probably have detected the damage in the ring.
The important point then is to take care of your bearings. Leave them in the box until you are ready to install them. Handle and care for them just as you would your precision measuring equipment. They are manufactured to that level of accuracy.

Spare Parts

The spares storeroom is often the orphan child of the building maintenance group. Spares may be kept in an outbuilding that is only opened when someone needs to get a replacement part. I've seen many cold, damp, somewhat dusty brand new parts that have been sitting on the shelf in an unattended storeroom, waiting to be used.

But, would you be surprised to know that bearings in the box have a shelf life, dependent on how they are stored. After manufacturing, they are packed with a preservative to protect the rolling elements, cages and rings. Depending on the conditions this preservative will degrade and adversely affect the bearing condition and it's life.

To maximize the service life of new bearings, SKF recommends the following basic housekeeping practices:
Store bearings flat, in a vibration-free, dry area with a cool, steady temperature.
Control and limit the relative humidity of the storage area as follows:
75% at 20 degrees C (68 F)
60% at 22 degrees C (72 F)
50% at 25 degrees C (77 F)
Keep bearings in their original unopened packages until immediately prior to mounting to prevent the ingress of contaminants and corrosion.

Bearings that are not stored in their original packaging should be well protected against corrosion and contaminants.

If you think you are not getting the service life expected from your bearings, go take a look at how they, and any equipment that has bearings installed, are being protected while they are in storage.

Stored Machines

A number of events that occur in our lives seem to repeat themselves on a somewhat regular schedule. For example a large number of maintenance personnel appear to cycle into and out of their jobs on a three to four year cycle. I base this on the types of bearing questions we receive at our online maintenance help site. For a specific example it has been noted that as apparently new people move into maintenance, they ask questions about how to properly store machines for the short or long term. Then these types of questions fade out for three or four years and then start appearing again. Because I've been in this business for 30 years, I've seen this question come and go for several cycles.

So, let's review machine storage for the next cycle.

When any bearing is mounted and stored where it is supporting the weight of the rotor, damage is likely to occur to the rings of the bearing. When the rotor remains stationary for periods of time, outside vibrations cause the rolling elements to minutely impact the rings. The rolling elements are made of a harder steel and make point contact with the rings, resulting in a condition known as false brinelling. The small impact damage to the ring is so small that they can only be seen with a microscope. But, when the machine is placed in service this very small area, which is usually in the load zone, slowly spreads into a visible spall leading to noise and vibration. Then the questions are asked as to why this new machine failed in such a short time. A quick solution to this problem is to set up a procedure to rotate the shaft every few weeks, thus changing the impact points.

The best example of poor storage was seen a few years ago in a dairy operation, which used a large number of pump/motor sets. In our inspection of the facility we toured the machinery storage facility filled with pumps and motors sitting on the concrete floor. By chance, while we were there, a train passed by the rear of the warehouse. We could feel the vibrations through the soles of our shoes. And the pumps and motors were receiving the same vibrations but whereas they didn't bother us, the vibrations were surely brinelling the bearings.

We recommended they find another space to store the equipment or at least get them off the floor and onto an absorbent pad. Every time a train passed by the bearings were being impacted as if someone were hitting them with a small hammer.

It's often the small things that make a difference between a smooth operation and an operation that spends lots of time doing unnecessary work. When you finish your coffee, go take a look and see how your equipment is stored and note the environmental conditions.

The Cost of Air

How often do you think about the air you breath? Unless it is dirty, hot, cold or muggy air, most of us don't give it a thought, just keep going and keep breathing without any active thinking on our part.

Now, how about the air in your plant? No, not the air you breathe but the compressed air that is often used throughout the plant for many purposes all the way from maintenance shops to production lines. Studies have shown that a plant system that is not well maintained can lose 10 to 30 percent of the compressed air in the network.

One of the most common points for air leaks is user equipment that is taken out of service but not removed. Each machine should have a shutoff valve on the incoming supply line. If it was not done during installation it is a cost saving addition. An even better solution is to install an automatic shutoff valve so that when the machine is not operating, the supply is shut off.

The best way to find air leaks is with an ultrasonic leak detector. Ultrasonic sound cannot be heard by the human ear but if the lines are physically traced with a leak detector, the detector will alert you to the presence of a leak.

Is it worth all this trouble? Based on published U.S. cost charts, a 1/16th diameter leak on a 100-psi system will lose 6.5 CFM. At an average cost of $0.18 per 1000 cu.ft., the annual cost of this small leak is $1,091.

Little leaks add up and one can see that there _is_ a cost of air. If your plant doesn't have an ultrasonic leak detector, there are maintenance services that will provide both the equipment and the men to operate it.

The Little Things Matter

There is a proverb that starts off with "For the want of a nail, the shoe was lost. For the want of a shoe, the horse was lost" the proverb, or one of many variations, ends with loss of a nation all for want of one nail.

In a parallel maintenance world we have many similar occasions. I worked with a large candy manufacturer who didn't feel it was necessary to monitor vibration levels on any electric motor less than 10 HP. His reasoning was that they are cheap, easy to handle and if one failed it was no big deal. What he didn't consider was that down on the production floor there was a 5 HP control motor that his whole production depended on. And of course one day it failed and the plant lost over 200,000 pounds of candy in various stages of a continuous flow process. For the loss of the motor, the line was loss, for the loss of the line the oven was lost, etc., etc.

Closer to home, the production output was a precision machine that makes bearing rings. During break-in tests prior to shipment, this $250,000 machine exhibited a high over all vibration on the grinding spindle. So the spindle was changed. But there were no changes in the vibration levels. Then all the pulleys were replaced. Same results, no change in the vibration levels. Next the pulley drive motor and belts were changed. The results? After three expensive days with a delayed shipment, the vibration amplitudes did not meet quality standards required for shipment.

We were called in to investigate. Instead of using an overall vibration measurement via a vib pen, we used a frequency analyzer to determine the source of the vibration. Much to everyone's surprise, the prime frequency matched the drive motor rotation speed; the motor appeared to be out of balance. Amid much huffing and puffing by the manufacturer he declared that it was a high priced precision motor so it couldn't be out of balance.

We asked them to remove the pulleys to run the motor solo for another check. As they removed the drive pulley, we noticed that it was a clamp on style and did not require a key for installation. Seeing that, we asked where the 1/2 key was that would fill the keyway cut into the motor shaft. More huffing and puffing, "It doesn't require a key can't you see that it very securely clamps onto the shaft".

We then pointed out to them that the motor was designed and balanced to operate with a key installed. If there was no key that was not a problem but a 1/2 key <u>must</u> be installed to keep the rotor in balance. All was quiet while a 1/2 key was installed, the pulley replaced and the machine energized.

With smiles all around, the machine was completely in standards for vibration and was cleared for shipment.

Now for the proverb again, version XX, "For the loss of the 1/2 key, the machine was lost, for the loss of the machine, the profit was lost, for the loss of a profit, **YOU** might lose your job.

Little things matter.

Tunnel Boring Machines, Drill Head Bearing

Tunnel boring machines have been in the news lately, the one drilling under Seattle is stuck, Las Vegas just finished drilling out under Lake Mead to bring in water to the city and New York City is in the process of building a new drilled water line from northern New York. A number of years ago these same types of machines drilled the tunnels under the channel between England and France. (The machines are still there, they drilled side tunnels and sealed them inside)

SKF had the opportunity to work with one of these machines when the nuclear waste depository tunnels were being dug to store waste deep inside a mountain on the nuclear test site.

The drill head on the machine is 28 ft. in diameter and supports the bearing of almost the same diameter. Overall the machine is several hundred feet long although there are variations depending on the job. It is not a drilling machine as with a pointed screw, the entire 28 ft. diameter of the drilling plate is forced against the mine face. Attached in this plate are a large number of pie plate shaped rotating disks of hardened steel. The forces applied to the mine face with the rotation of the drill head and the rotation of the hardened steel disks causes the face to shatter and drop to a conveyor system to exit the shaft site.

SKF was asked to determine the condition of this drill face bearing, in place, a mile inside the mountain. The machines cannot reverse, they only go forward. Therefore if there was a problem with the bearing a "room" would have to be hand dug around the drill face to be able to remove the bearing.

Using enveloped acceleration technology, the bearing condition was monitored using a portable Microlog with a 100 mv/g accelerometer. Data was collected on the bearing frame as the drill face was rotating.

It nearly caused a panic when the data was analyzed later that day in the office. There was a very clear indication of a fault in the outer race of the bearing, (BPFO). The fault was clearly seen in both the FFT and the intervals in the time spectrum. The bearing manufacturer was contacted and the observation was explained to them.

Their answer solved the problem. After the bearing was completed, annealing wires were wrapped completely around the bearing so that the heat would anneal the steel and increase the hardness. To prevent the starting end of the wiring from overlapping the terminal end and overheating a small section making it brittle, a "small" gap was left between the two ends of the heating wires. This left a small area of the bearing with a "soft" spot. When the rolling elements passed over the spot, the sensitivity of enveloped acceleration process used by the data collector was able to detect the impacts, which were then correctly interpreted as a fault. Technically it was a "fault", but it fact was a design feature that had no effect on the operation of the bearing.

As a follow up: When the tunnel was completed, in a horseshoe configuration, the machine was parked outside the exit portal and advertised for sale. A proposed buyer was located and they asked SKF to again check the bearing prior to the purchase. After a two-day trip to get to the site and the hassle of the security clearances, we arrived and prepared to do the testing. We asked them to turn the machine on; they said they couldn't because it had no power.

We packed up and went home, the machine may still be parked there in the desert.

Unexpected Consequences

A small manufacturing plant specialized in precision shafts. The company had been in business for a number of years in the same location. As business improved, the decision was made to expand the facilities including the storage yard. The yard was used to store uncut shafts and other materials used at the plant.

After the expansion the percentage of rejects of the precision shafts increased to an unacceptable level. After their normal trouble shooting was completed and the reject rate had not decreased we were asked to come in and see if the problem was in some way related to either a bearing problem or a vibration problem. After completing our testing, we could find no bearing or vibration source for their problem.

The problem was rather straightforward. The uncut shafts were placed in a precision lathe and turned to the desired diameters with the usual shoulders and groves as required. While in the lathe all the dimensions were verified and with the current problems, verified by a QC person. Everyone agreed that the shafts were exactly as designed. Since the shafts were ordered on an "as needed" basis, they were immediately shipped to the customer.

However when the shafts were received by the customers, some of them would be rejected by their QC for being slightly bowed or out of round as measured in microns.

When a problem develops for no obvious reason the first thing to look for is "what has changed." The expansion of the plant was the first thing on the list. Over a period of several days we followed the shafts from the storage shed through the plant, during manufacturing and to the shipping department. Nothing was happening to the shafts that could result in the dimensional discrepancies that were occurring.

Near the end of the second day, about the time we were ready to give up, the cause and effect were found. When the expansion took place, the uncut shaft covered storage area, was moved from the north side of the building to the west side of the property. This new storage area was also well covered but the sides did not extend all the way to the ground, leaving about a 4 ft. height exposed. And that was the problem.

Being located on the north side the building kept the uncut shaft storage area in the shade of the main building all day. At the new location, in the afternoon, some of the uncut shafts were exposed to direct sunlight for their full length. The sunlight resulted in uneven radial heating of the shaft. When this unevenly heated shaft was placed in the lathe, it was turned to specifications and approved by QC. When it cooled to an even temperature, the dimensions were changed and it was no longer acceptable. Shafts from other locations not exposed to sunlight, or work done on cloudy days, continued to pass inspection because they were not heated. Only these random warm shafts proved to be a problem.

The solution was simple, bring tomorrow's production run into the main building the day before they were to be worked. They could have completely enclosed the storage building but bringing them inside the day before was the simplest and cheapest solution.
Problem solved.

How's the Weather?

When I first started doing condition monitoring an important part of the standard equipment was a 50-foot extension cord. Required because all the equipment had to be plugged into line voltage and invariably, there wasn't a permanent socket where you needed one. Hence the extension cord with multiple sockets or maybe even two cords.

Today life has become much easier. Not only has electronic miniaturization greatly reduced the size and weight of the equipment, our 30-pound "portable" data collectors now weight 3 pounds and our primary equipment is powered by batteries. With the advances in memory storage, functions such as printing no longer have to be done in the field and can be accomplished in the comfort of an office environment. Life is good!

Good until some unforeseen event occurs that changes everything.

On the shores of the Gulf of Mexico south of Houston, Texas is a ship building company that was constructing an oil-drilling platform to be used in the gulf. It was the middle of March and we had been asked to check the bearings in the anchor chain drives. These drives had been installed early in the construction of the platform and had been exposed to the elements for over a year, unpowered, so they had not been rotated the entire time. Before accepting the rig, the owners wanted the bearings checked for their condition without the expense of disassembling the drive equipment.

So we loaded our equipment and drove to the nearby motel. Beautiful clear day so we checked in, locked up the equipment in the car and had a peaceful nights rest planning on an early start the next morning. Much to our surprise the next morning it was COLD. There wasn't frost but it didn't feel like the middle of March anymore.

Reporting to the rig the next morning we set up our equipment and were ready to begin testing. But, within a few minutes the equipment shut down with a low battery warning. That wasn't a problem, we had spare batteries in the car but they also operated but a short time and displayed a low battery warning. Do you remember what happens to batteries when they get cold? They go to sleep and don't want to wake up. We were fortunate that we had been doing a customer demo at our previous stop and were carrying a battery charger and 120-volt line power was available.

It wasn't the most professional looking exercise, take some data, change the batteries, install the charged spares, put the old batteries in the charger, take a little more data, swap the batteries, take some more data, swap the batteries. We got through the morning but it could have been prevented if we had been aware a cold front was coming through and had taken our equipment inside the room instead of leaving it in the car trunk.

Someone said we have to learn from the mistakes of others because we can't live long enough to make them all ourselves. So be aware, don't leave your battery-powered equipment in the trunk of the car in the winter. Or the summer either for that matter, heat is also hard on batteries and electronic equipment.

What Does That Warranty Cost?

It is common practice to provide a warranty with new machines. If it fails or does not perform as advertised it will be repaired or replaced. Usually a condition added to maintain the warranty is that the machine has to be overhauled every 18 or 24 months. Consequently equipment that is operating in good working order is taken out of service and overhauled per the warranty specifications, work performed based strictly on the calendar.

The question is, is this a good maintenance practice? The rules are followed and everyone is generally happy. It may be a good maintenance practice but is it a good financial practice? In general, no, it is not. This does not apply to safety equipment but for general purpose equipment the U.S. Navy has accepted the old saying, "If it ain't broke, don't fix it".

Of course overhaul costs will vary with the size and complexity of the machine but for the purpose of comparing cost we shall assume it cost $30,000 to overhaul Machine A and it is processed every 18 months. New bearings, seals, gaskets, hoses, whatever the warranty calls for, year in and year out. Simple math says that after 6 years maintenance and 4 overhauls, they will have spent $120,000 to maintain the machine and keep the warranty current.

Across the road, in another company, machine overhauls are carried out based on machinery conditions. Over the 18 month period the machine performance is measured and compared with historical or manufacturer's guidelines. At the end of 18 months if the machine is performing satisfactorily it remains in service and the monitoring continues. At the end of 24 months it was decided to overhaul the machine just to be sure it is O.K. This 24-month cycle continues for 6 years and the maintenance cost are $90,000 and the second company has saved $30,000.

This scenario is hypothetical but in real life was followed by a major university in the maintenance of their parallel chillers. The chillers were operated so that they had the same amount of operating time. Chiller A was overhauled on an 18-month schedule and Chiller B was overhauled when condition monitoring indicated that maintenance was required. At the time we were there they showed a cost savings of slightly more than $100,000 on Chiller B and the decision had been made to overhaul both machines only when the need was indicated by the monitoring data.

And one other consideration. Maintenance studies have shown that after maintenance, 11% of the time the machine will be out of service within 6 weeks to correct a maintenance error or replace a new part that did not perform as designed. Such things as a pinched seal, bad gasket, cocked bearings, or some other malfunction.

"If it ain't broke, don't fix it".

Your Brand New Machine Is Broken!

We work for one of the major bearing manufacturing companies. The equipment they use to manufacture bearings comes from all over the world since different companies provide tools that can perform some functions better than others. Needless to say these machines can be very expensive and even if that were not a consideration, when they are out of service then production schedules are thrown awry.

One of the many stages that a bearing goes through during manufacturing is the honing of the rings. As with any smoothing operation, the process starts with a light pressure and then increases the pressure for the final finish with an extremely fine man made stone. After the plant was commissioned everything went well with just the normal glitches of a major startup.

After several months, the cutting fluid positive displacement pumps began to have internal failures, the internals were completely destroyed, there was no repair, they had to be replaced. After the third pump failed we were asked to do an analysis to determine the cause. As we observed the operation it was readily apparent when the fluid pressure was suddenly increased, there was a very loud impact sound and the entire machine shook.

When the pressure valve cycled a water hammer was occurring which instantly put a large load on the pistons of the pump. In fact the loads were high enough that the steel screws were pulled out of the aluminum housing. There were shreds of aluminum in the threads of the screws. Our recommendation was that they install a surge tank in the line by adding a vertical "T" bar with both ends of the "T" capped. The trapped air would compress and reduce the fluid forces when the valve cycled.

Because the machines were new and still under warranty, the manager did not want to make any modifications so we recommended that he contact the manufacturer who was located in Italy.

We later learned that when the contact was made, the first comment by the machine manufacturer was "Oh, don't you have the latest modification?" The honing machines were a new design and other users were also having pump problems. The design team had arrived at the same conclusion we did and had designed a surge chamber to be installed in the fluid line from the pump. Their chamber was more professional looking than our proposed "T" bar but it solved the problem after installation.

The moral of the story is that even brand new machines will break if an unexpected design problem enters the picture. The parallel story to this is the major car manufacturer who experienced the exhaust pipes breaking off their new model cars. Turned out that where the pipe brackets were mounted allowed the free end of the pipe to vibrate at its resonance frequency when the car was driven about 60-65 MPH. This flexing of the pipe induced metal fatigue and at some point it would break off. The bracket was moved to a different location dampening out the resonance and the problem was solved.

Vibration Analysis

Basic Vibration Analysis

There are numerous mechanical conditions that can be observed using portable vibration data collectors such as the SKF Microlog. The following series of blogs are the major faults and conditions one may find.

Balance:

As a general rule whenever a machine is vibrating, the first impulse is to assume the rotating element is out of balance. The quickest way to verify this is to collect an FFT in the velocity spectrum.

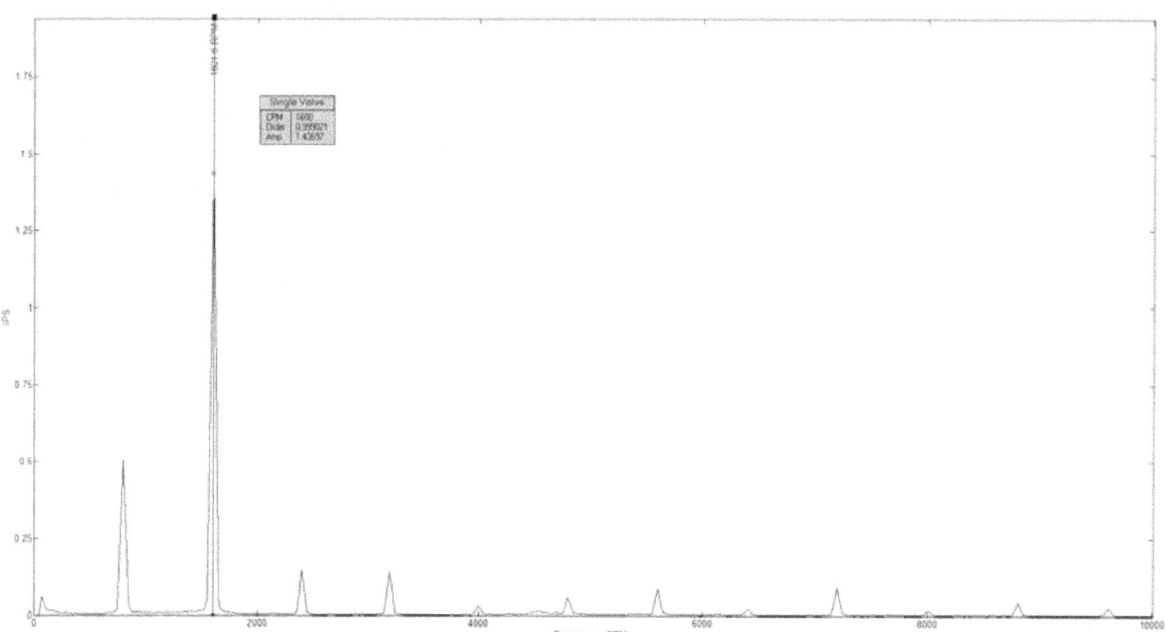

With the cursor on the high peak, the data box tell us that this fan is rotating at 1600 RPM and the imbalance is 1.4 IPS. An extreme amplitude for any machine and completely unacceptable. The small spike of energy at 800 RPM is probably from two components rubbing together because energy at 1/2 rotation speed is an indication of a rub. The other spikes are insignificant and can be ignored.

This fan was the suction fan on a bag house. A bag had come loose and hung up on one blade of the fan, putting it out of balance. The bag was removed and the following spectrum taken.

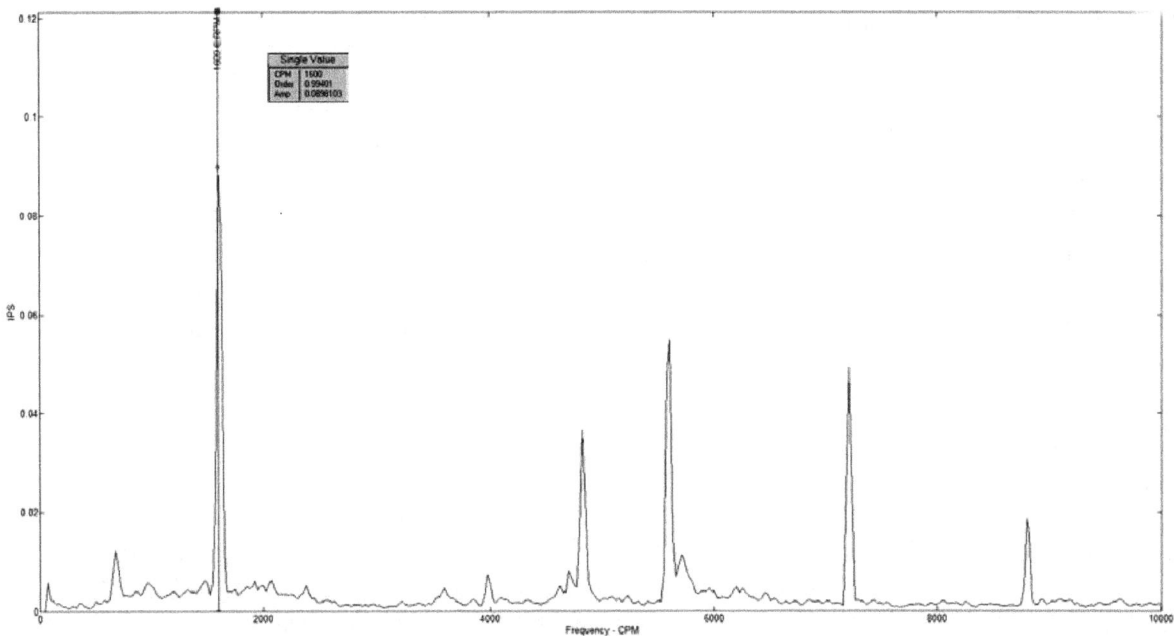

What do you see different between the two spectrums.

Warning, this is a trick question to illustrate an important point. Most data collectors can be set to "Auto Range". That means that it will change the values in the "Y" axis so that the spectrum will fill the screen, as this one does. Visually one would say, "There's hardly any change". However the amplitude box tells us that the RPM is still 1600 but the amplitude is now 0.09 IPS, a 15-fold decrease in amplitude that is very acceptable for continuing operations.

Characteristics of an FFT Indicating Misalignment

Because misalignment induces a "push-pull" force for each shaft rotation, two actions per rotation, there will be a display of energy at twice running speed, (2X). Because these forces are parallel to the shaft, the best location to collect vibration data when checking for misalignment will be on the ends of either the driver or driven unit i.e., in an axial direction. Since there are rarely perfectly aligned units, it is not unusual to see some indications at 2X. However, concern should be aroused if the 2X amplitude is more than 50% of the 1X amplitude when the 1X is greater than 0.15 inches per second. There is also a 1X factor although it is masked by any imbalance that is present and can only be documented by noting the reduction in the 1X signal after the unit is in alignment.

The phase relationship across the coupling will show that the motor and driven unit is 180 degrees different, plus or minus approximately 30 degrees. This difference is measured between the inboard bearing of the motor and the inboard bearing of the driven unit. This difference is caused by the push-pull action that for one-half of the rotation is pulling the two pieces of equipment together and then is pushing them apart on the second half of the rotation so that they are always moving in opposite directions, i.e., out of phase.

There are two types of misalignment, parallel and angular. In parallel misalignment, the two shafts are parallel with each other but not in the same horizontal plane. In angular misalignment, the two shafts may be in the same plane but are at an angle to each other. Although angular misalignment is the more common problem, there is often a combination of both problems when the machine base is not properly prepared. Parallel misalignment will be more prominent in the radial measurement and angular in the axial direction. Both will have the 180-degree out of phase condition across the coupling.

Field Example
Figure 1 1191 RPM Pump/Motor Before Alignment

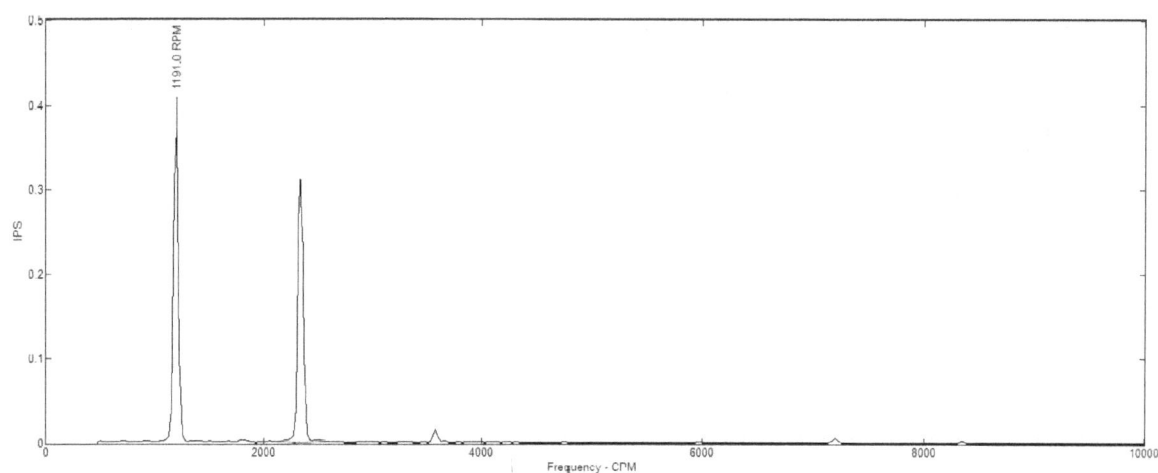

Figure 2 1191 RPM/Motor After Alignment.

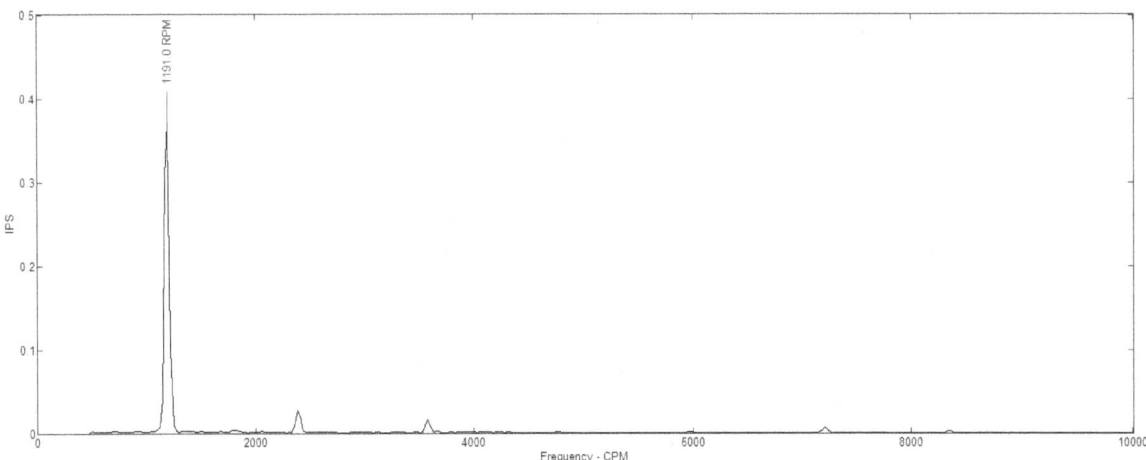

This water injection pump was reported to be vibrating and the FFT displayed Figure 1. Maintenance was requested to perform an alignment. The face-to-face misalignment was found to be 0.010". Although the alignment has been corrected, the 1X amplitude is too high indicating an out of balance condition that needs to be corrected.

Another aspect of misalignment that is often overlooked is the additional energy required to operate equipment in a misaligned condition. Field-testing has determined that the additional power required to operate a misaligned pump/motor set is 2.5-3.0% higher than a properly set up unit. Using data from a large refinery with hundreds of motor/pump sets, the savings for just one machine was calculated in the following manner.

Three phase KW = $\dfrac{(volts)(amperes)(P.F.)(1.732)}{1000}$

(KW difference)($/KW)(7200 hrs/yr) = $ Annual Savings
7200 hrs/yr is assuming 6 day week - 50 wks/yr

EXAMPLE: Initial amperage = 27 after alignment = 25
Initial KW = $\dfrac{(575)(27)(.8)(1.732)}{1000}$ = 21.5

After Align = $\dfrac{(575)(25)(.8)(1.732)}{1000}$ = 19.9 KW

21.5-19.9=1.6 KW Difference

(1.6KW)($.06/KWH)(7200 Hrs/yr) = $691.20/yr

As noted in previous blogs, the major result of misalignment is damaged bearings. Many coupling manufacturers advertise that their couplings can absorb the stress of misalignment. This may be true, but the bearings cannot. The energy used by misalignment is converted to heat and vibration, energy that will eventually damage the bearings no matter what coupling is installed.

Costs of Resonance

The cost of resonance was well documented at a refinery where there were two identical 300 HP vertical pumps that were periodically swapped out during operations. This facility kept very good maintenance cost records and after 10 years Pump A had incurred $200,000 more in maintenance costs vs. Pump B. It was well known that Pump A had a problem, it was nicknamed "Old Shaky", but no one was able to devise a solution to the vibration.

A quick way to check for resonance is to install a temporary brace and see if the vibration decreases. A small hydraulic jack was placed between the wall and the motor and pressure applied. It was observed on the data collector that with each stroke of the jack, the vibration decreased until it was within an acceptable range.
With these results, a bump test was performed to verify the suspected resonance.
The data from a bump test is in two parts, the time domain and the frequency domain spectrums. The machine is struck and the input is seen on the screen.

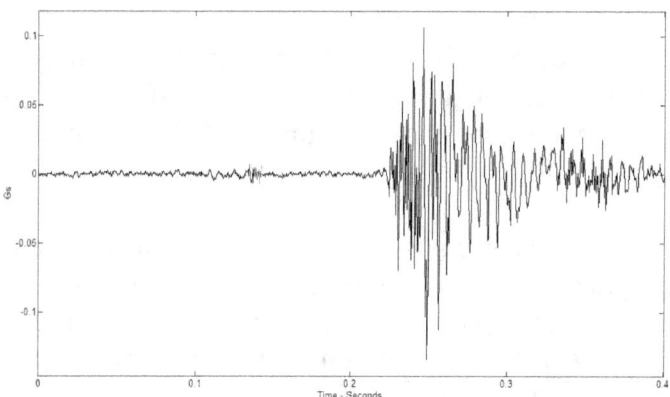

The importance of this time spectrum is to insure the energy induced into the machine has completely decayed by the right edge of the plot. Because this energy is used to develop the frequency spectrum, if the decay is not complete, the collected energy that is used to derive the frequencies will be incomplete and in error.

The data collector then processes this energy information and derives the frequency spectrum.

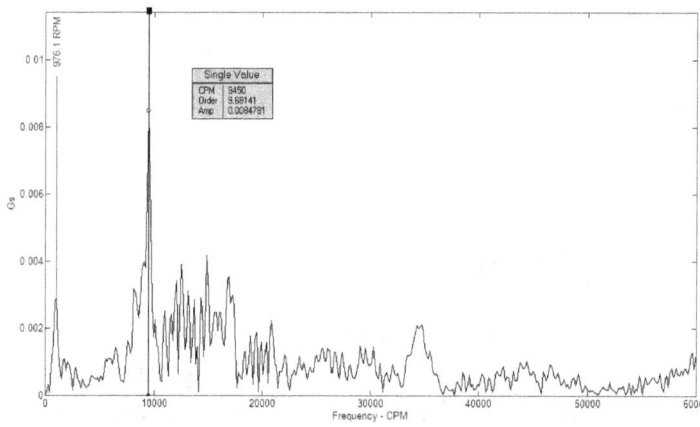

Although there are many frequencies present, the critical one is at 976 CPM because the pump operated at 1000 RPM. Rotational speeds should not be within 20% of a resonance frequency.

Comparing other data from the "B" pump it was found that the foundation of the "A" pump was not level and in plumb as it should be. With a cocked base, the vertical pump was not properly supported and vulnerable to resonance.

The faulty foundation was removed and replaced. When the pump was placed back in service, "Old Shaky" was no more and the vibration levels were normal. But there was a clear record that the resonance problem had cost the plant $200,000 in labor and materials over the past 10 years.

Basic Frequency Analysis--Spectrums--

Frequency analysis involves learning a new language. If you have ever looked at a frequency spectrum and your first question was, "What do all those wiggly lines mean?" then you are just like everyone else the first time they were presented with a data spectrum.

Some definitions and background:

Frequency spectrums are presented in a format called a Fast Fourier Transform or FFT. Fourier was a French mathematician who conceived this visual method of presenting vibration signals. This record displays the frequency on the "X" axis (the horizontal axis) and amplitude on "Y" axis (the vertical axis). Frequency can be visualized as how fast an object is moving back and forth or up and down and is expressed in cycles/minute (CPM) or cycles/sec (Hz). The faster it is moving, the higher the frequency. Amplitude is expressed in displacement, (how far it moves, mils, inches), velocity (how fast it moves from point to point, IPS), acceleration (the rate of change in the movement, G's) and enveloped acceleration, (gE). All these measurements can be expressed in Metric or English units.

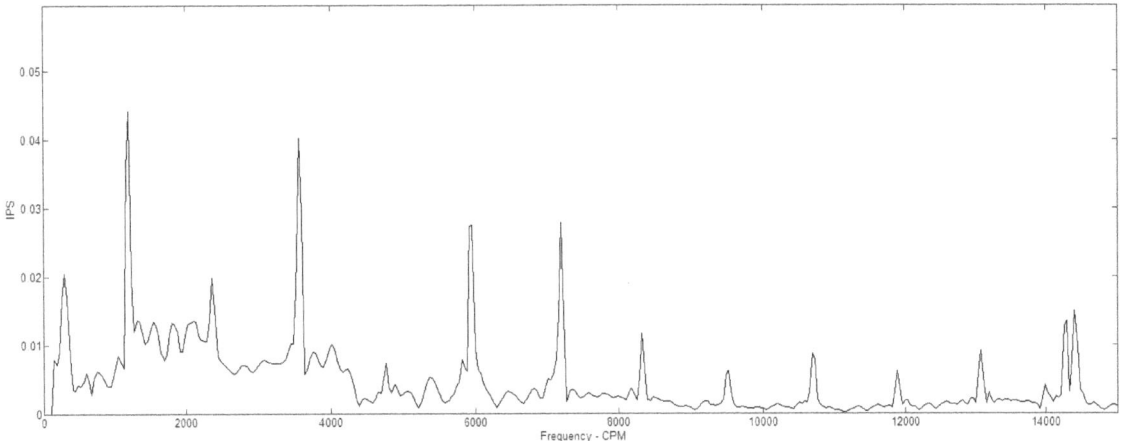

This is a typical FFT, with the ("Y" axis) amplitude measured in inches per second (IPS) and the ("X" axis) frequency measured in cycles per minute (CPM).

The data collector would supply you with additional information such as the overall amplitude, the total energy measured from 0 to 15,000 CPM. An FFT breaks down the total into the amount of energy at various frequencies. On the data collector is a courser, which enables the user to place the courser on an individual energy spike and determine how much energy is present at that specific frequency.

For example if the machine being examined rotated at 1200 RPM then the first large spike of energy is at approximately 1200 CPM and has an amplitude of approximately 0.042 IPS. When energy is seen at rotation speed it predominately indicates an out of balance condition. But no routine machines are perfectly balanced so this low amount of energy is no cause for alarm. Machine balance tolerances are established in a number of ways, manufacturers, users, ASME, etc.

The other outstanding feature of this spectrum is the presence of harmonics. A harmonic is defined as the presence of energy at equal frequencies apart. In this case, 1200, 2400, 3600, 4800 CPM, etc. When harmonics are a function of the rotation speed (1200 CPM) it is usually an indication of loose components i.e., bolts, foundations, bearing housings, etc.

There are always spikes of energy that are mysteries. Some are from other machines and some you never know where they come from; don't lose any sleep worrying about them.

Of course there are a lot more aspects of vibration analysis and if there is any interest shown I can continue writing in this vein.

How Much is Too Much?

A re-occurring question that pops up every 4 or 5 months concerns the amplitude of bearing faults and when should action be taken. As with any component installed in multiple systems, it is impossible to set absolute values but for general guidelines the following suggestions may be of help in your daily routine.

These values are based on data collected using <u>enveloped acceleration</u> as the metric. There is NO conversion formula to covert these values to any other measurement metric, velocity, displacement, or acceleration.

In a bearing, damage in the outer ring is the most common occurrence. In most cases the outer ring is always in the load zone whereas the rotating inner ring spreads the load throughout the ring. Of course there are exceptions where the inner ring is fixed and the outer ring rotates, just allow for this difference in your analysis, the vibration values remain the same.

Amplitude: 0 to 0.3 gE

Occasionally these low values may be seen on a new bearing. It is not common but it does happen depending on the manufacturer. If there is damage in the bearing it cannot be seen without magnification. There is no cause for alarm nor is any action required at this time. Note the occasion and verify the data on the next observation.

Aplitude: 0.3 to 0.6 gE

There is probably some visible damage such as a spall developing in the load zone. It is recommended that the monitoring of the bearing be increased such as instead of once a quarter, inspect once each month. It would be advisable to also notify maintenance that the bearing is degrading and will need repair in the near future.

Aplitude: 0.6 to 0.9 gE

Significant damage has occurred and can be seen during a physical inspection. This bearing should be changed at the next opportunity. If there are sideband frequencies of other components, cage or rollers, the machine should be taken out of service and the bearing changed.

Amplitude: 0.9 gE and higher

This bearing should be changed immediately. If operations says the machine must run, then the machine must be monitored daily. There is NO way to predict when a bearing will fail, never give a date that failure will occur. If failure occurs before that date, you lose. If failure occurs after that date, you lose. Just state that the bearing is damaged and failure can occur at any time.

The amplitudes of early bearing damage will trend upward in a fairly linear manner. As the damage increases, the slope of the trend will increase but at some point above 1.0 gE, the trend will go toward vertical and failure will occur.

Resonance

The root of the word resonance comes from a musical word, resound. If a musical reed that is tuned to a "C" note, for example, is placed next to another "C" note reed and the first reed is made to vibrate, it will cause the second reed to also "resound" and vibrate. In the mechanical world this same influence takes place when an out of balance rotor rotating at 1800 RPM, which would be vibrating at 1800 CPM, is mounted on a frame that has a natural frequency of 1800 CPM, the frame will also begin to vibrate. The motor and frame would then be said to be "in resonance." This example shows that resonance is not a vibration source, but the response to another vibration.

The frequency of a structure is influenced by its stiffness and mass. If these values are known, it is possible to calculate the natural or resonance frequency of a machine or structure. Another way is to perform a "bump" test and "ring" the structure by striking it and measuring the response with a frequency spectrum. A large bell is not rung by striking it with a straw. When a bump test is performed, the mass that is used as a hammer must be large enough to excite the structure, just as a heavy clapper rings the bell. Structures will have a different response when struck in a horizontal plane vs. a strike in the vertical plane because the stiffness will be different in each plane.

When structures are in a resonance condition a small vibration input, such as imbalance or misalignment, is amplified and in time may damage the structure. The natural frequency of the structure should not be within 20% of the rotation speed of the rotating element. So if 1000 RPM is the rotation speed, the structure should be designed to have no natural frequencies between 800 and 1200 CPM. There are only two ways to change the natural frequency of a structure, increase the mass to lower the response and stiffen the structure to raise it. Mass can be supplied by addition of bags of lead shot and added bracing will provide addition stiffness, both have been used successfully.

Although structures can be designed properly, the owner often gets into trouble by operating the machine at a speed not included in the original design. This has been the experience of many papermaking machines, which over the years have been slowly increased in speed to increase production. It is not unusual to find placards on operating panels with strict instructions on speed ranges that must not be engaged during operation when equipment is operated with variable speed drives. Also, it is possible with variable speed drives to operate at a speed, which excites an internal component such as bearings. If a variable speed drive is involved and the bearing life is abnormally short, there are cases where the bearings have been damaged by resonance. Since the bearings mass and stiffness cannot be changed, it is again necessary to block out a specific operating speed range.

The term resonance and natural frequency mean the same thing, it is the frequency determined by calculations or performing a bump test. A critical frequency is the frequency that if maintained will result in damage if the generation of that frequency is continued.

There Are Limits

If you have been in maintenance for very long then you have learned that not all things work out as you have planned. One comment we often hear is that the customer had been doing his vibration monitoring by the book just like he was trained to do and yet the machine failed with no warning.

The case at hand occurred when a customer in Asia contacted us with the complaint that a machine that he was monitoring once a month had failed 17 days after he had done his last monitoring. He included the FFT plot and amplitude trends and we concurred that there was no indication in the data that the machine was in any trouble. We asked for a teardown report for the cause of the failure.

The report stated that the root cause was a failure of the cage in the outboard bearing.

Rarely in the real world do we have equipment that always, every time, without fail, year after year, perform as designed. With vibration monitoring there will be occasions where the data doesn't give a warning of impending failure. Cage failures of bearings is one of these occasions.

The cage is generally the lightest weight of the 4 components in a bearing. Therefore, it generates the lowest amplitude signal and then the signal has to cross over three barriers; cage to ring, ring to housing, and housing to sensor. In addition, whereas a spall in a ring or rolling element will often develop over a period of weeks or months, the light weight cage, if damaged, will usually deteriorate rapidly, often in a matter of days.

The results is that you may have good data on the 1st of the month, as in this case, and by the 17th, something has damaged the cage and it has entered a failure mode which shuts down the machine. On cage damage the usual suspect is installation error that is initially undetectable. On rare occasions, with continuous monitoring, we have observed cage fault signals and have shut the machine down under controlled conditions and performed the necessary maintenance but the usual scenario is a shutdown after the machine emits a loud screeching noise. Two solutions for critical machinery is to increase the observation rate and the other is continuous monitoring, both of which have economic limitations.

So, as the title says, there are limits. Nothing is 100% sure except death and taxes. Therefore we work within the limits of our equipment and do the best job possible.

Time Domain Spectrums

On a previous blog we discussed the main features of an FFT. Another visual presentation used in vibration analysis is the time domain spectrum. It is often neglected but it is a valuable tool in analyzing repeating vibrations such as in a gearbox with a cracked tooth or a loose key in a keyway.

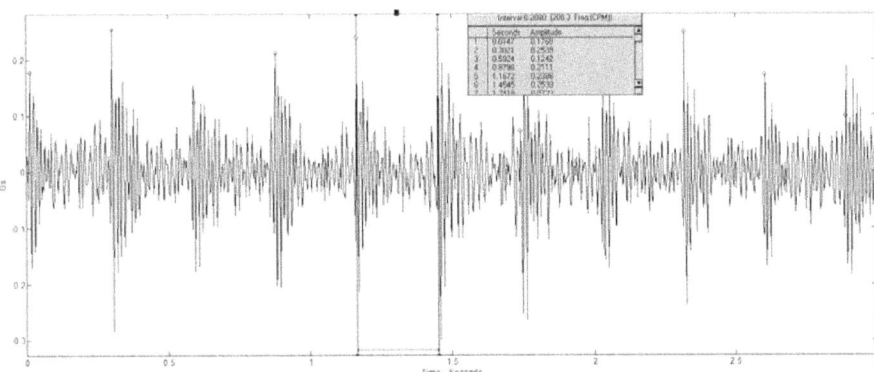

This is a typical time domain spectrum. The amplitude in on the "Y" axis and the time is shown on the "X" axis. This spectrum is from a gearbox with a broken tooth. It is important to remember that the data in the time domain is the same information that is seen in the FFT but it is seen as a "side view" of the FFT. If one could go to the side of this page and look at the edge and it were expanded to a greater thickness, there would be the FFT. It was that Frenchman, Mr. Fourier that figured out mathematically how to do this transformation.

In this example, which gear has the broken tooth? Software allows us to set a cursor on one spike of energy and to set a second cursor on the next cursor. (left or right). The software then calculates the time difference, 0.0147 and the frequency of this occurrence 208.3 CPM. By knowing the input RPM (900 RPM) and the gear ratios it was calculated that the output gear was rotating at 208.3 RPM. It was important to know which gear was damaged because in this case they did not have a spare 208-RPM gear. So before they shut the unit down, they ordered a new gear and had it on hand at shutdown.

To display the same vibration data in an FFT, the FFT is shown below.

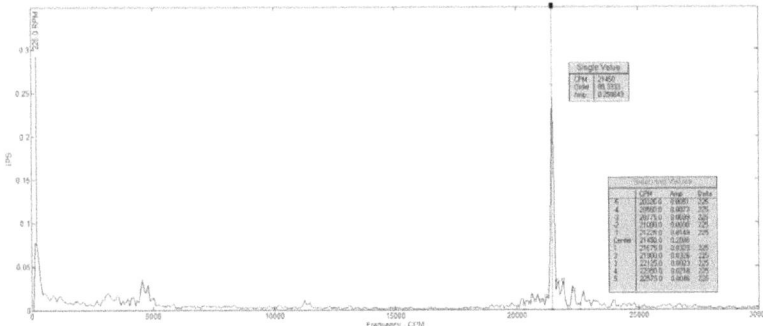

This FFT shows the gear mesh frequency of 21,450 CPM and if one measures the delta frequency of the sidebands, the software calculates it to be 225 CPM. The difference between 208 CPM from the time spectrum and 225 CPM on the FFT is due to the difference in resolution between the two spectrums. But a high-resolution spectrum takes a longer time to collect and the difference comes down to "close enough". Also the overall amplitudes of the FFT and the time spectrum will usually be different. The time domain overall is a true peak, the amplitude of the highest peak whereas the FFT is an average amplitude of all the data collected. If there is a great difference in the amplitudes, that is a clue that something is giving the unit a good impact at a random rate while the FFT averages a single input with the others and the overall average is much less than the single impact.

Vibration Analysis Tools, gE

The prime tool for vibration analysis is the portable Microlog series of data collectors in conjunction with the included dedicated software designed for specific purposes. In addition to the portable data collectors, permanently mounted equipment with the same internal processors and software are available for use where it is not feasible to use portable equipment.

For rolling element bearings analysis, the ultimate software to use in conjunction with the Microlog, is Enveloped Acceleration, (gE) This unique, patented software was developed by SKF specifically for bearing analysis and has been in use, with upgrades, for 20 years. It is unique because in contrast to other vibration measurement methods, it cannot be converted to any other measurement such as IPS, G's or Displacement.

It does have some characteristics that the user must be aware of and incorporate in the analysis. For example, when using velocity as a measurement, a reading of 0.35 IPS would be considered rough over a wide range of speeds, 600 RPM to 5,000 RPM. With enveloped acceleration the rotation speed of the bearing must be considered in the final analysis.

During initial testing of the software, a bearing used in a DC motor was deliberately damaged in the outer ring with a carbide tipped tool. When the motor was operated over it's speed range from 50 RPM to 3600 RPM, multiple vibration readings were taken on several run ups and run downs of the motor with no change to the bearing.

The following chart was developed plotting the RPM in the "X" axis and the amplitude is plotted in the "Y" axis. It can clearly be seen that as the speed increases, the amplitude for the same amount of bearing damage will result in an increase in amplitude. In this chart the amplitude at 50 RPM is 0.004 gE increasing to 1.8 gE at 3,600 RPM.

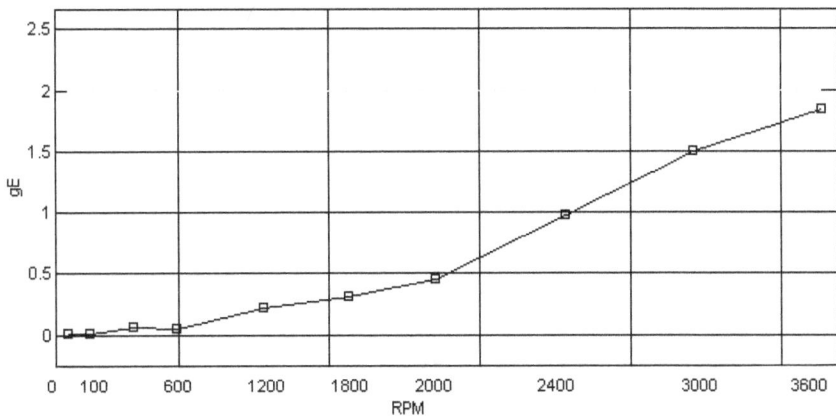

Therefore if you were looking at this bearing at 3,600 RPM it would immediately be seen that a 1.8 gE reading indicated sever damage and the bearing should be changed. However, based on previous experience using velocity measurements, a reading 0.1 at 600 RPM may not be considered sever when in fact, the same amount of damage is present. This may be more clearly understood if one looks at the formula, F=MA. In this case the "M", mass, remains the same so it can be set to equal "1" leaving us with F=1A. As the speed increases, the centrifugal force "F", increases, leading to a direct increase in the measured acceleration signal, which is processed by the Microlog and presented as enveloped acceleration.

You must take this feature of enveloped acceleration in consideration when you are making your damage evaluations.

Vibration Analysis: A Maintenance Information Tool

In any endeavor the more information we have about a situation, the better will be the decisions made regarding that situation. In maintenance, the more information we have about the condition of a machine, then the better we can maintain that machine.

Using a portable vibration data collector, such as the SKF CMXA80, vibration analysis allows the maintenance technician to take a "look" inside a machine while it is in normal operation and understand how the various components are operating. A day of training will provide the technician with the BASIC skills to evaluate, for example, the condition of the bearings, both plain and rolling element, evaluate the alignment of the driver and driven unit, confirm the balance condition of the rotating elements and check for looseness between machine parts. Additional training will further enhance the operator's skills in other areas of analysis.

A musical conductor can distinguish the various instruments in an orchestra by the different frequencies generated by each instrument. Vibration analysis uses the same technique. When placed on a machine, an accelerometer captures a large range of frequencies and the data collector processes and displays them in a format known as a Fast Fourier Transform (FFT). Each different part or component of a machine generates a different frequency. For example, damage in the outer ring of a bearing will generate a unique frequency that is different from the frequency generated by damage on the inner ring, the rolling elements or the cage. The point to remember is that frequency identifies the source of the vibration. On the FFT display, the frequency scale is shown on the horizontal, "X" axis.

Next, the question of how much damage has occurred? The amplitude, which is displayed in the vertical, "Y" axis, is an indication of the amount of damage present or if displacement is measured, how much the machine is moving. Depending on the type of analysis being conducted, an estimate may be made of light, medium or serious damage. Here the point to remember is that the FFT amplitude provides an estimate of damage.

In sum, frequency identifies the source of the problem; amplitude tells you how bad it is. It is important to remember that you never give a date when the machine will fail. If it fails before that date, you lose. If it is still operating on that date, you lose because it didn't fail. It is best to just advise, management that the machine condition is normal, has a problem developing or is in trouble.

An example of how this information can be used: You have four identical machines and you have time to only overhaul one machine. Which one do you work on? By evaluating the vibration levels and the condition of the various components, you can grade the machines from 1 to 4 and you know to overhaul the high-ranking machine, #3.

Vibration analysis is an excellent maintenance evaluation tool.

What Causes My Machines To Vibrate?

We actually received a phone call and that was the question the caller asked. We explained that there were entire books written to answer that question and they were often subdivided to the point where another book was written on each chapter. But to just give him an idea, here is a short list of the primary causes of machinery vibration. This is our list; someone else will have a different outlook and a different list.

At the head of most lists you will find unbalance, defined as a condition where the center of mass is not in the center of rotation. Think of your car tire. If there is more rubber in one location then this shifts the center of mass in that direction. The condition is corrected by placing an equal mass 180 degrees opposite and at the same distance from the center of rotation or removing some of the excess mass. Even when unbalance is not the problem when someone places a hand on the machine and feels a vibration, many will say, "Something is out of balance".

Although there is some disagreement in the order, the next most common cause of vibration is misalignment of the driver and the driven unit, i.e., a motor and pump. Compounding the problem is that there are multiple types of misalignment. In general if the two shafts are not aligned both vertically and horizontally the unit is misaligned. As a sidebar, misalignment is the cause of up to 70% of bearing failures. Coupling manufacturers say that their couplings will handle misalignment, which is fine, but the bearings can't handle the undesigned loads condition for a long-term life.

On most lists mechanical looseness is near the top of the list. It is just as it says; there are loose bolts or cracked welds that allow the parts of the machine to move independently of each other when they should move in unison. Experience has shown that looseness usually follows hurry up work or the mechanic that says "That's good enough". It often isn't!

Finally on this short list we could include resonance. Every machine, and its components, has a resonance frequency. As a rotating mass increases in rotation speed it will pass through one or more resonance frequencies (RPM's) It is totally unacceptable to attempt to operate a machine in a resonance condition. Any machine that is set up for variable speed, as with a variable speed drive (VFD), usually has a placard warning the operator of what speed or speeds to avoid. Ignoring this warning will lead to the destruction of the machine. Actually, resonance isn't a vibration source, it is an amplifier that increases an existing vibration, increasing as the "resonance or natural frequency" of the machine is approached.

That's the start of some reasons machines vibrate, it's not the final authority but if you have a machine that is shaking, look at these causes first and in 90% of the cases you'll find your source of the vibration.

www.ingramcontent.com/pod-product-compliance
Lightning Source LLC
Chambersburg PA
CBHW080931170526
45158CB00008B/2242